CAMBRIDGE LIBRARY COLLECTION

Books of enduring scholarly value

History of Medicine

It is sobering to realise that as recently as the year in which On the Origin of Species was published, learned opinion was that diseases such as typhus and cholera were spread by a 'miasma', and suggestions that doctors should wash their hands before examining patients were greeted with mockery by the profession. The Cambridge Library Collection reissues milestone publications in the history of Western medicine as well as studies of other medical traditions. Its coverage ranges from Galen on anatomical procedures to Florence Nightingale's common-sense advice to nurses, and includes early research into genetics and mental health, colonial reports on tropical diseases, documents on public health and military medicine, and publications on spa culture and medicinal plants.

Great Artists and Great Anatomists

A gifted yet controversial anatomical teacher, Robert Knox (1791–1862) published this remarkable study in 1852. It explores the influence of anatomy on evolutionary theories and fine art respectively. The first part of the work discusses the lives and scientific insights of the eminent French naturalists Georges Cuvier (1769–1832) and Étienne Geoffroy Saint-Hilaire (1772–1844). Rejecting the explanations offered by natural theology, Knox maintains that descriptive anatomy can give answers to questions surrounding the origin and development of life in the natural world. The latter part of the book is concerned with the relation that anatomy bears to fine art, specifically the painting and sculpture of the Italian Renaissance. Entering the debate about the importance of anatomical knowledge in art, Knox focuses on 'the immortal trio' of Leonardo da Vinci, Michelangelo and Raphael. Henry Lonsdale's sympathetic biography of Knox has also been reissued in this series.

Cambridge University Press has long been a pioneer in the reissuing of out-of-print titles from its own backlist, producing digital reprints of books that are still sought after by scholars and students but could not be reprinted economically using traditional technology. The Cambridge Library Collection extends this activity to a wider range of books which are still of importance to researchers and professionals, either for the source material they contain, or as landmarks in the history of their academic discipline.

Drawing from the world-renowned collections in the Cambridge University Library and other partner libraries, and guided by the advice of experts in each subject area, Cambridge University Press is using state-of-the-art scanning machines in its own Printing House to capture the content of each book selected for inclusion. The files are processed to give a consistently clear, crisp image, and the books finished to the high quality standard for which the Press is recognised around the world. The latest print-on-demand technology ensures that the books will remain available indefinitely, and that orders for single or multiple copies can quickly be supplied.

The Cambridge Library Collection brings back to life books of enduring scholarly value (including out-of-copyright works originally issued by other publishers) across a wide range of disciplines in the humanities and social sciences and in science and technology.

Great Artists and Great Anatomists

A Biographical and Philosophical Study

Robert Knox

CAMBRIDGE
UNIVERSITY PRESS

University Printing House, Cambridge, CB2 8BS, United Kingdom

Published in the United States of America by Cambridge University Press, New York

Cambridge University Press is part of the University of Cambridge.
It furthers the University's mission by disseminating knowledge in the pursuit of
education, learning and research at the highest international levels of excellence.

www.cambridge.org
Information on this title: www.cambridge.org/9781108065283

This edition first published 1852
This digitally printed version 2013

ISBN 978-1-108-06528-3 Paperback

GREAT ARTISTS
AND GREAT ANATOMISTS.

GREAT ARTISTS AND GREAT ANATOMISTS;

A BIOGRAPHICAL AND PHILOSOPHICAL STUDY.

BY

R. KNOX, M.D., F.R.S.E.,

LECTURER ON ANATOMY, AND CORRESPONDING MEMBER OF THE "ACADÉMIE NATIONALE" OF FRANCE.

LONDON:
JOHN VAN VOORST, PATERNOSTER ROW.

M.DCCC.LII.

LONDON:
Printed by SAMUEL BENTLEY and Co.,
Bangor House, Shoe Lane.

ADVERTISEMENT.

THIS work is composed of two parallel Biographies. The first comprises the Life and Labours of George Cuvier and Geoffroy (St. Hilaire), the men who have most contributed to the development of the true relation of Anatomy to the Science of Living Beings. In the second part the reader will find a brief history of the relation of Anatomy to the Fine Arts. In the parallel biographies of Leonardo, Angelo, and Raphael, the Author is convinced that ample materials exist for the decision of the long-protracted controversy in respect of the relation of Anatomy to the Arts of Sculpture and Design. He is at the same time well aware that long prior to the great men whose lives he has here sketched, others existed with minds equal if not superior to them, but who, from pursuing other studies and other aims than the political game of life, constitute, notwithstanding, an epoch or era, less brilliant, less fiery, perhaps more durable, than the epochs of Cæsar, of Alexander, and Napoleon. Such was Aristotle, and such the men who carved the Venus, the Laocoon, and the Apollo. But of the lives of these latter, little or nothing is known: they left no writings explanatory of the Canons of Art; the works of the great masters in painting have disappeared, whilst the matchless sculptures alone remain to attest a power of mind and a civilization which we scarcely yet comprehend. Although the Canons of Art must have been well understood by them as their discoverers, yet it is certain, that, however admirably they appreciated the relation of Anatomy to Art, they had never studied Anatomy. To some this will appear a paradox: but if those who think so will favour me with a perusal of this work, they will, I hope, find the paradox solved. The true relation of Anatomy to Science was perfectly understood by Aristotle. Such at least was the opinion of Cuvier himself, the greatest anatomist—*Descriptive Anatomist*—of any age. He preferred, as more exact, Aristotle's description of the anatomy of the elephant to that of Daubenton, his own immediate predecessor, be it remarked; for Mertrud was nobody in science. Nevertheless, the author remains of the opinion, that prior to Bichat, exact Descriptive Anatomy, the greatest of all elements in the study of living beings, was unknown to Aristotle and to the world; whilst, in respect of philosophy, whatever the divine genius of the Greek may have grasped, a demonstration of his theory could not be given, so long as the anatomy of man and the human embryo was unknown to him. It remained then for Cuvier, Geoffroy, and Leonardo to test the true relation of Anatomy to Science, Philosophy and Art. The conjectures of the Ancients they converted into theories; they formed the era in which they lived. The object of this work then is threefold, 1st. To establish the exact relation of Descriptive Anatomy to the science of the animal organic world, as it now is and as it once existed. In the life and labours of George Cuvier, as he views them, the Author finds this relation fully made out. Before Cuvier appeared, geology was a farce, a subject of ridicule; cosmogony a myth; the history of creation a tissue of error and absurdities. 2nd. To trace Transcendental Anatomy to its essence, and to show, in the life and labours of Geoffroy (St. Hilaire), that the philosophy of the creation of animals is explicable only by Descriptive Anatomy. 3rd. To discover, if possible, in the life and labours of the immortal artist who painted the "Cena," and of his great rivals, Angelo and Raphael, the true relation of Descriptive Anatomy to Art.

Other matters are no doubt discussed in these scientific Biographies, for of such this work really consists. It were impossible, for example, to overlook the fact, that there are men whose lives form an epoch in man's history; whose lives form, in fact, the history of the period in which they live. Such was Newton in respect of science; such was Aristotle; and, politically, such were Cæsar, Alexander, and Napoleon; what these men were in respect of the brute masses of men, those I now speak of were to the thinking world. Yet they wielded but one element of knowledge—Anatomy — Descriptive Anatomy — a science not yet fully understood in Britain.

Throughout this work, by the term Science is meant a knowledge of the living organic world, man of course included, in relation to the existing circumambient media; its relation as it now appears to us, in time and in space, and the relation which each great natural section or grouping bears to all others. But in this definition the author ventures to include also the positive knowledge we have obtained through the discoveries of the immortal Cuvier, of the pre-existing organic forms, known by the name of Fossil Remains. The demonstrations of the relations of the groups of animals and plants comprising these organic worlds, apply equally, but less rigorously, than in the case of the now existing organic worlds, to these ante-historic organic configurations of life, owing to the destruction of nearly all the soft and perishable materials. The terms "former, or ancient world," "past creations," "successive organic worlds," are terms to which no definite meaning can be attached, there being in reality no such things. All these great results, as regards the progress of human knowledge, flow from the application of an element of thought, first discovered by Bichat, who applied it only to man; re-discovered by Cuvier, who applied it to the animal world in its *entirety*. It reacted on all other branches of human knowledge, by bestowing on the minds of men an illimitable expansibility of thought, which is power, as proved by its immediately, and for ever, altering the character of human reason. Even in Cuvier's time its application, by Geoffroy, to the philosophy of organic beings, startled the scientific and thinking world.

By the term philosophy is meant, throughout this memoir, the result of the application of exact anatomy to the embryonic structures, and of the knowledge so acquired, to the theory of the origin of forms; from this, with the aid of the Cuvierian researches, resulted a *demonstration* of the unity of all organic beings from the beginning to the end—the past, present, and future—the discovery of the true relation of that which has been to that which is, and, without doubt, to that which is to be; a new cosmogony in fact, the direct result of that science, geology and paleontology, which the world owes to Cuvier; the restoration, in fact, of the history of creation to the subordination of those secondary laws which regulate all material things—a bringing, for the first time, within the pale of strict science a department of human knowledge which Aristotle and Lucretius, Leibnitz, Pascal, and Newton, had in vain endeavoured to effect.

Lastly, by Art is meant, the "Fine Arts," that is, Sculpture and Painting. Art, thus defined and circumscribed, the author places among the diviner arts, which eminently distinguish man from the mere animal. All men are not destined merely to wield a sabre and to feed a pig. The connexion of Art, with some knowledge on the part of the artist of the interior structure of man, has never been questioned; but the precise relation which Anatomy bears to Art, has not yet been, in the Author's opinion, determined. In the life and labours of Leonardo, Angelo, and Raphael, he offers a solution, or an attempt at a solution, of this difficult question. Fortuitous circumstances, highly favourable to the testing this great question, brought into contact the three great masters of modern times,—Leonardo, Angelo, Raphael. On all three, Nature had bestowed a divine genius, matchless hands, an intense longing for the perfect, a power to perceive and strongly to admire the truth. They re-discovered the beautiful and the perfect in Art; their minds were universal. But to each she had also given an individuality of character, which, by enabling them to look at the external world in the bright mirror of their own reflection, furnishes to the historian the means of solving the important question I have already alluded to,—What is the relation of Science to Art? The discovery of this relation seems to have been reserved for Leonardo.

GREAT ARTISTS

AND

GREAT ANATOMISTS.

PART I.

SECTION I.

George Cuvier,

HIS LIFE AND LABOURS.

THE author of the following discourses has been long persuaded that the true relation of anatomy to science, philosophy, and art, has not yet received from thinking men the attention it merits. To supply a deficiency which he believes to exist in the history of the progress of the human mind from error to truth, the discovery of which he presumes to be the only rational end of human existence, he ventures to offer to the public, in a biography of George Cuvier and

Etienne Geoffroy, the views which, after much consideration, he has adopted on the relation of anatomy to science and philosophy; and in those of Leonardo, Angelo, and Raphael, the relation of anatomy to art,—to the divine arts of painting and sculpture.

In the lives of Cuvier, Geoffroy, and Leonardo, including those of Raphael and Angelo, the author fancies he sees the development of the one great principle which has led to such glorious results. Assuming exact descriptive anatomy to be the basis of all zoological knowledge—using the term in its most extensive signification—the author unfolds in the life of Cuvier the application of descriptive anatomy to zoology, living and extinct; in other words, to the science of organic beings. In the labours of Geoffroy he sees the application of anatomy, transcendental and abstract, but still essentially descriptive, to philosophy; in that of Leonardo, it is not difficult to trace the application which that great master and discoverer made of true descriptive anatomy to art. The author submits his views to the candid and the unprejudiced of all races.

That Bichat was the founder and discoverer

of true descriptive anatomy, the author is ready to admit. But Bichat confined his method, or at least his followers did, to man's structure and to practical science—to the arts, in fact, of surgery and medicine. Cuvier, cognizant, no doubt, of what Bichat had done, extended his mode of research to all other animals, and thus he made of zoology a science. But, above all, by this method, by this new element of knowledge, was he enabled to read the true character of the fossil remains of all epochs, and, for the first time, to present man with a "History of the Earth," not founded on fables, but on facts.

From the beginning to the end of his career, the nature of his inquiries was either mistaken or misrepresented in Britain. His philosophic discourse on the changes which the surface of the globe and its living inhabitants had undergone "in time," was presented to the British public as a "Theory of the Earth," a "New Theory of the Earth!" He gave us instead a "History of the Earth," whereon to build a theory. It was the old mistake of Bolingbroke, who called history, philosophy teaching by examples. But history is not philosophy. Cuvier gave us a history of the world; the philosophy of that history he never attempted.

Prior to Cuvier, geology, paleontology, cosmogony, had really no existence; what passed for such were dreams. Before Geoffroy, or rather before Goethe, the origin of life, the phases and metamorphoses of living beings, from the period when this orb commenced its wild but measured career through space, had been wholly misunderstood; a slavish terror of free inquiry hung over men's minds, dark as the pall of night.

When Napoleon was first consul; when law and equity, though based on despotism and discipline, resumed their sway in France; when life and property came once more to be, in a sense, respected, there appeared, in the capital of that great country, two young men of humble prospects, parentage, and means. The path they followed was eminently obscure, unobtrusive, and retired. They were naturalists! They belonged to a class of men who investigate the external characters of animals, with a view to discover how far they differ from each other; in what a dog, for example, differs from a cat; a gull from an oyster-catcher; a sparrow from a linnet; a crocodile from an alligator; a bee from a wasp. Their definitions of animals, plants, and minerals, are generally diverting, often ludicrous. They take the trouble to prove, that man is not a monkey, and never was a monkey, which is more, however, than they can vouch for. They give you rules and definitions of character, to enable you to distinguish an oyster from a muscle, a whelk from a periwinkle.

In this endeavour to create terminology, the terminology, too, of beings for which man has no sympathies, into a science, they forgot and forget the principles of all education, and the nature of the human mind. Ask the schoolboy how he distinguishes the bee from the wasp; the ass from the horse; the red deer from the fallow deer; and, be assured, that he will laugh at you. Could he answer you philosophically, his reply, probably, would be—"For what purpose has Nature gifted me with powers of observation, through my senses? How do I distinguish a chair from a table? Is it by definition? Trouts from perches? perches from carp? Is it by definition? For what purpose has Nature given me that practical tact, which deals with natural appearances, placed beyond all your definitions, furnishing me with that practical knowledge never to be acquired in books or schools?" Or ask the indolent Bosjesman, as he listlessly gazes with you over the Great Karroo, or scans with his telescopic sight the beauteous plains of the Koonap, or the slopes and tangled rocky dells of the Annatolo, how he came to know the name of all animals around him, from the majestic lion to the harmless blue bok; of every creeping thing, serpents and lizards, iguanas, scorpions? Would not his answer be the same as the school-boy? "In this land I was born, and brought up from my earliest years; instinct taught me to discriminate animals, plants, rocks, soils. The various antelopes I distinguish from each other, even by their slightest movements, by their attitude, when colour and shape are lost in the distance; by their numbers and grouping, when they move not; by the character of the ground and pasture on which they roam." This would be his answer could he reason with you; and did he know that in certain great and civilized communities, there are schools and colleges to which men have been appointed to mislead and misdirect the youthful mind; to teach him to substitute for his own tact and powers of observation, a barren terminology; to accept a definition for actual observation through the senses; words for ideas; the untutored savage, the uncivilized man, would, no doubt, modestly recommend his more energetic white brother to return to nature and to truth.

That methodical and systematic works, on what is called natural history, are necessary for the advancement of science, and the unfolding the truth, I do not deny. No one is more alive to the necessity and advantage, in a scientific point of view, of such studies and such pursuits. It is to the mode in which they are taught that I object. In respect of the practical *utility* of such pursuits, the common sense of mankind has already pronounced its verdict, in unmistakable terms. As regards the vegetable world, a doubt has never been expressed by any race or nation. Against the necessity of a profound study of, and an extensive acquaintance with Nature's works, as the animal, vegetable, and mineral productions of the globe are usually called, men, practical men, of all races and nations, have pronounced a verdict; they have declared such knowledge to be useless and vain. In fact, to the most of Nature's living productions, man is the direct antagonist. Over all that is beautiful in her wild Flora, he drives the ruthless plough. Her thousands and tens of thousands of plants, ever varied, ever beautiful, he roots out and destroys, as filthy, useless weeds. For him, the tree which yields not a plank or a spar, or a useful beam, is a curse and an encumbrance on that soil, on which he can afford room for nothing which ministers not to his wants. With man, savage or civilized, all is utility! To the wide expansive ocean he grudges its limits, calling it the unprofitable, the untillable sea. Minerals of exquisite beauty he tramples under foot, converts into *metal* for his roads, or hews into blocks for walls and bridges.

Utility again! it is all utility with man, conceal it as you may. To the animal kingdom of Nature he is equally antagonistic. On the discovery of a new land, man's first object is to destroy nearly every living creature which Nature put there. If it will not, or cannot come within the pale of domesticity, for which Nature, it seems, did not intend it, the animal so offending becomes an object of pursuit; man devotes it to destruction. Nature and her works are nothing to him. And should any compunctious feelings arrest for a moment his hand, staying the wide-spread desolation springing up around him, he is warned by unerring instinct of the tenure on which he holds his position on the globe—destroy and live, spare and perish.

Whilst so engaged, he stumbles on the fossil remains of a former world, I was about to say; but this, though stereotyped, is an incorrect expression, and simply misleads; it should be, "the organic remains and the inorganic products which existed in ages by-gone;" of ages, countless in number as the sand of the shore.

To the uninquisitive, the utilitarian man, the man of to-day, these dead and marrowless bones are objects of no value, saving in as far perhaps as reduced to powder, they may again manure his fields. They teach him nothing: satisfied with what he has been told by the pedant, the ill-informed historian, what he gathers from fabulous tales of oriental myths, he conjectures, if he thinks at all, these bones to be vestiges of ancient history; what a history! the remains of animals which lived in the time of Cæsar, or it may be of Moses.

Quarries were dug in the olden time; Mount Athos was tunnelled by Xerxes; a canal connected the Nilotic waters for many centuries with the Red Sea; and the crust of the globe had been dissected by the metallurgist and engineer. Fossil remains had been seen by millions of men, ere Cuvier appeared. But man would not, or could not, see the truth. All things swam in the chaotic deluge of the Roman poet; shell-fish rested on the tops of mountains, and fishes took refuge amongst trees! The human mind, oppressed by conventionalism, was unequal to describe simply "the anatomy of man." At last appeared the man, gifted with the *desire to know the unknown;* the anatomist.

To the quasi-philosophic men of his day, practitioners of medicine and surgery, profoundly ignorant of the structure of that animal they practised on, Bichat offered the "Descriptive Anatomy of Man;" Cuvier went further.

"These bones, which you conjecture to have belonged to elephants and crocodiles, and horses and men, did not belong to any such animals. The exact anatomy of animals which now live teaches me, that, provided *species are not convertible into each other* (an hypothesis he mistook for a theory), these bones are the remains of an organic world which has ceased to be. Suddenly, and as if by magic, the obscuring veil, the thick pall of ignorance, drops from before human sight; the mist disperses from hill and valley; a vast and wonderful land, redundant with life, exhibiting ever-varied, gigantic, and grotesque forms, is spread out to the gaze of the admiring observer. That observer was George Cuvier. Still what he saw was but an image, a phantom of the past. His view was backwards into remote antiquity, whilst yet the world was in its infancy. Occupied with facts and details, that is, history,—eschewing principles, that is, philosophy,—his view, even of the past, was limited and confined. That past he did not fully comprehend, or rather he avoided admitting that he did; of the future he said nothing. Simultaneous with him arose others, who valued facts merely as leading to principles; of these, Goethe and Geoffroy may be considered the type and the leaders. Other illustrious names must be conjoined to these. They did not discover the transcendental in anatomy, but they collected the facts in support of its principle, and they applied them to the history of organic life, not merely as it is now, but as it has been, and as it may be in futurity. Thus two men, and two modes of thought, overturned all existing knowledge, all existing chronology, all human history. Descriptive anatomy, which Cuvier and his followers called comparative anatomy, in his hands overturned all existing cosmogonies: the transcendental went further; it developed the great plan of the creation of living forms; the scheme of Nature. It unfolded the secondary laws by which the transformations are made, the metamorphoses out of which variety springs from unity: the natural history of creation was for the first time explained to man.

The subject, then, which I purpose handling, resolves itself into two distinct parts; the results, namely, of true descriptive anatomy, on human knowledge; secondly, the effects of the transcendental, which is but a form of descriptive anatomy on the human mind.

They cannot well be separated from each other; without the facts derived from the descriptive anatomy of individual forms, the transcendental theory were a mere hypothesis, without proofs, unsupported by facts. Placed together for mutual support they then became irresistible, and could we *formule* the doctrine as simply as the divine author of the "Principia," it would take its place side by side with the theory of gravitation. The day is not far distant when this must happen. To explain how and in what order these remarkable events occurred, it is necessary to go back to the period when Cuvier and Geoffroy were yet young men, and to inquire into the state of Natural History, as it is called, at the period when these original thinkers commenced their bright career.

Carl Linnè, the prince of all classifiers, of all methodical naturalists, leaving behind him in this walk all other observers, offered to scientific men and to the world, his "Systema Naturæ," a work beyond all praise. The organic and inorganic worlds he classified and arranged. His grouping was at once natural and artificial. It was a master-piece of formulism, and, although in a direction differing much from that of Aristotle, yet equalled and even exceeded the "Historia Animalium" in its results; the end of both being the same, namely, the classification of all animals, plants, and minerals, with this double view: first, their precise discrimination; secondly, their arrangement according to their natural affinities. To the great name of Linnè, who had already made natural history, as he viewed it, fashionable throughout the civilized world, were soon added those of Buffon and Werner.

Buffon by the charms of description, Werner by his earnest reasoning, bestowed, the former on zoology, the latter on mineralogy, an interest they never before possessed, and never can again enjoy. The "external character" system or method attained its maximum of reputation, and was in the zenith of its glory when Cuvier and Geoffroy appeared in Paris. It was their destiny first to improve, extend, support, next to overthrow the system they previously looked up to; the theories of Werner received their final refutation from Hutton.

Before dismissing the era of the mere formulist (Cuvier also was a formulist in a sense), the man of external characters, from the stage of history, let me here do him the justice to say, that in so far as a *practical knowledge*, all but constantly exercised on things in which man takes no interest, goes, the system of Linnè is the only true, the only practical one. The zebra, the quagga, the ass, are not distinguished from their co-gener, the horse, by their skeletons or internal anatomy, but by their *external characters*. Nay, what is more, by their internal anatomy we do not readily discriminate them from each other. The robe, the external covering, the surface of the animal or plant intended to come into contact with the air, and with human vision, is infinitely the most characteristic part of every animal. It is that by which Nature has distinguished every species of animal and plant from all others. It can never be overlooked. It constitutes natural history, properly so called, in as far as regards the practical discrimination of one animal from another; but it fails, in toto, as a basis for the exact philosophical classification of zoology; another element of science, another art must be brought into play—descriptive anatomy. It fails also when used as a basis for the philosophy of zoology, another and a higher range of inquiry must be resorted to; the interior must be examined with other views than the mere comparison of one species of animal with another; it must now be compared with man as he is and as he was, with animals as we now see them and as we know they were when the world was yet in its infancy; lastly, the fully formed specialized animal with the embryonic stages through which he passes, or has to pass, in other words the transcendental in anatomy, must be sought for, and when found applied to the history of creation.

The men who have contributed most in bringing out these great results, are the two persons whose life and labours I am about to chronicle; Cuvier, namely, the descriptive anatomist *par excellence;* Geoffroy, the transcendentalist. Like Linnè and Buffon, it was their destiny to give to natural history an interest it never had before and never will have again.

Thinking man is anxious only after principles; these once established, withdraw from the facts establishing them all human interest. At the commencement of his career, Cuvier made the descriptive anatomy of the lower animals interesting to man. The facts he unfolded in respect of the mollusca, surprised and pleased. He extended this interest even to the anatomy of dogs and cats, of rats and mice, of moles and rabbits, by the beauty of his descriptions. But with him all interest ceased, and when attempted to be revived by a mere anatomist of much greater power than Cuvier, the attempt failed, and Meckel's great work on what is usually called comparative anatomy, but what is really the descriptive anatomy of the lower animals, fell dead from the press. But I anticipate a singular fact in the history of science and philosophy, and shall proceed at once with the life of that man to whom mankind is indebted for the greatest of all discoveries, leaving to follow that of Geoffroy, whose view of Nature, when sufficiently supported by facts and further researches, and philosophically formuled, will take its place with the discovery of the laws of gravitation.

GEORGE CUVIER, the first of all descriptive anatomists, and the scientific man who first, after Aristotle, applied the art of anatomy to general science, was born on the 23rd of August, 1769, at Montbeliard, a small and originally a German town, but long since incorporated within the French territories. He was a native of Wurtemburg, a German in fact, and not a Frenchman in any sense of the term, saving a political one. The family came originally from a village of the Jura, bearing the same name, of Swiss origin therefore, and a native of the country which gave birth to Agassiz. In personal appearance he much resembled a Dane, or North German, to which race he really belonged. Cuvier then was a German, a man of the German race, an adopted son of France, but not a Celtic man, not a Frenchman. In character he was in fact the antithesis of their race, and how he assorted and consorted with them it is difficult to say. Calm, systematic, a lover of the most perfect order, methodical beyond all men I have ever seen, collective and accumulative in a scientific point of view; his destinies called him to play a grand part in the midst of a non-accumulative race, a race with whom order is the exception, disorder the rule. But his place was in the Academy, into which neither demagogues nor priests can enter. Around him sat La Place, Arago, Gay Lussac, Humboldt, Ampere, Lamarck, Geoffroy. This was his security, these his coadjutors, this the audience which Cuvier, the Saxon, and therefore the Protestant, habitually addressed. It was whilst conversing with him one day in his library, which opened into the Museum of Comparative Anatomy, a museum which he formed, that the full value of his position forced itself upon me. This was, I think, during the winter of 1821 or '22. A memoir had been discussed a day or two before at the Academy: I remarked to him that the views advocated in that memoir could not fail to be adopted by all unprejudiced men (*hommes sans préjugés*) in France. "And how many men *sans préjugés* may there be in France?" was his reply.

"There must," I said, "be many, there must be thousands."

"Reduce the number to forty and you will be nearer the truth," was the remarkable observation of my illustrious friend.

I mused and thought. Napoleon was as

good as dead to the world. Louis the Fat and Gross festered and rotted in the Thuileries; the priests were gradually acquiring their lost influence. Still intellectual France was comparatively free, and Cuvier and Geoffroy, Humboldt and La Place, could still live and think. How different must have been the lot of Cuvier had fate cast his nativity in Britain; there he must have lived and died, "alike to fortune and to fame unknown." Poor, and therefore despised, what could the simple minded pedagogue (for in his youth he was a tutor) have effected against Oxford, Cambridge, and the cliques of London? What part could he have played in the weekly farce at Somerset House? His anatomical labours and views held in the most sovereign contempt, as Hunter's were by the meanest country apothecary; sneered at by the metropolitan physician and surgeon; frowned down by the theologian, as dangerous and leading to scepticism, he must have quitted England, or turning his vast intellect to some profitable pursuit, and abandoning science for ever, taught mathematics to boys, chemistry to the apothecary's apprentice, or the anatomy of the parts of the body, concerned in surgical operations, to medical students. This was the state of England and of science in England during the greater part of Cuvier's career.*

John Hunter, it is true, had lived and died, leaving his museum in the hands of those to whom it was, and still, in some measure, remains, a sealed book. But Hunter's researches were directed towards a totally different object than were those of Cuvier; and I do not mean, therefore, to compare them here. Each inquired after truth by his own path. Cuvier was the descriptive anatomist *par excellence;* Hunter was also quite equal to this; but he had other views. He was the physiological anatomist, unsurpassed, unequalled. A greater genius by far than Cuvier, he yet effected nothing compared with the laborious German. In the grand qualities of the human mind, those qualities which distinguish man from the brute; a desire to

* Descriptive anatomy, the great instrument of discovery in organic science, was quite unknown in Britain until 1815; and I preserve as a curiosity and a proof of this fact, a work called "The London Dissector," the standard book of the profession in London, and the type of the nation's mind in respect of the science of anatomy. I have made some more extended remarks on this point in note I., which will be found at the conclusion of the work.

know the unknown; a love of the perfect; an aiming at the universal; in these qualities Cuvier and Hunter agreed: but to discover new and unobserved phenomena, and to detect new relations in phenomena already observed; this faculty, that is genius, belonged eminently to Hunter.*

From various sources, and especially from the work of Mrs. Lee (Bowditch), a lady most favourably placed for acquiring an intimate knowledge of Cuvier and his family, we learn that his father was a person of slender means, and in no way remarkable. At school, George Cuvier showed great facility in learning all that was required of him, combined with a talent for method and order, which his kind-hearted biographer thinks indicated the dawning talent of the legislator. But Cuvier never was a legislator, in any sense of the term; he was a scientific, not a political man. Yet I am aware that, during the reign of Louis the Fat and of the miserable Charles who succeeded him, Cuvier was weak enough to fancy himself important in the administration of public affairs. It was quite a delusion. He fancied himself, also,

* See note II.

an orator; but he was not. He was a clear and methodical lecturer; a most polished and fascinating writer; but that is all. Returning from college to Montbeliard, he was necessitated to leave that, for want of means to pursue his studies, accepting the humble situation of tutor in a wealthy family. Chance sent him to Caen, in Normandy; this was in July, 1788, at which time he could only have been about nineteen years of age. From 1791 to 1794 he was still in Normandy, near the sea, without books, and fortunately without teachers. But Nature and the "Systema Naturæ" of Carl Linnè were both before him. He thus commenced to observe for himself. About this time, it is said, he began to think of comparing *fossil remains* with the now existing living world.

Sketching rapidly the life of the man, as distinct from his scientific career, I may observe, that, by the malevolence of the master of the Gymnase, in which he received his elementary education, he was sent to Stutgard instead of Tubingen; this influenced his views in life. He acquired the German language at Stutgard; at the Académie Caroline he dissected with Kiesmeyer, adopting, however, the study of administration as the future pursuit of life. This he soon abandoned for natural history, or rather for the pursuit of zoology. In 1796 the National Institute of France was created, and Cuvier was called to Paris in 1795, chiefly by the instrumentality of that man, whom he afterwards overshadowed,—Geoffroy. He declined accompanying Napoleon to Egypt, though invited to do so. Appointed to assist Mertrud, lecturer on comparative anatomy at the Jardin des Plantes, he prepared and published his first great work in 1800, the "Leçons d'Anatomie Comparée." It produced no sensation in England, where, indeed, its object, owing to the character of the prevailing race, was wholly misunderstood.

In 1800, Mertrud died, and was succeeded by Cuvier. A new career was now opened to him, which he entered on with unsurpassed energy. He founded and created the great Museum of Comparative Anatomy, still in Paris; prosecuted his anatomical studies; revised the organic kingdom; and prepared his grand work on the "Ossemens Fossiles;" that work which was to revolutionize all human knowledge, save the merely mechanical. In

1802 he received other Government appointments, such as inspector-general of education, with instructions to establish Lycées in thirty towns. He was now elected perpetual secretary of natural sciences in the Institute. In 1811 appeared the "Ossemens Fossiles," and Cuvier reached at once, by universal consent, the highest possible reputation as a scientific man. From this period, until his death on the 12th of May, 1832, he never ceased for an instant the pursuit of science and of truth.

But he advanced not; and by the influence of his great name and position, became an obstructor of science. Latterly he resisted all attempts to theorize; and, as a leader of a numerous body of partisans of all nations, he became the bitter and uncompromising enemy of Geoffroy and the transcendentalists. He did his utmost to crush these men, and to drive them from the Academy. Sufficient for him it seemed to be, that he had established the *great fact*, that the species of animals now alive, and forming the organic world since human history commenced, differ essentially, specifically, and generically, from those whose remains, fossilized, we now discover in various parts of the world.

He called this merely a fact! and so it is, no doubt. Cuvier called his great discovery a fact. It is a fact so far as it goes; the most extraordinary fact ever discovered by man; but it is, as we shall perceive, a discovery rather than a fact, admitting of no modification. By this discovery Cuvier upset all existing cosmogony, natural history (if it merited the name), geology; but to convert his discovery into a fact, applicable to all ages, to science, involved several hypotheses, which he at first admitted, afterwards rejected. The eternal fixity of species was one of these, and this included the non-convertibility of one animal into another by any secondary cause whatever; by climate, by domesticity, by time, by geological epochs, or cataclysms; lastly, by the eternal laws of development, forming an intrinsic attribute of living matter. Cuvier was scarcely dead, when my illustrious friend, De Blainville, so connected the living rhinoceros with the extinct fossil genera by a series of individuals, as to leave little or no doubt of the identity of the genus, at least; the identity of the present with the past. The mammoth of Cuvier, and his mastodon, genera as he fancied so distinct from the elephant of the present world, were proved to be connected therewith by a chain of species *occurring in time*, so resembling each other, so little characteristic as distinct species, that the idea of species began to fade from human thoughts. It was this great law of transition, of metamorphosis, which alarmed Cuvier in his later years, although it ought not to have done so —Nature's transitions of organic life in time and circumstance; the formation of all *living forms* from one living essence. His dislike to see in the living world, past and present, one animal instead of many, was caused simply by a dread of its touching that reputation, which he knew the world based on his having proved the contrary.

In whatever way the transitions are effected, they are purely the results of secondary causes; to abandon this view is to abandon human reason. Transitions of organic beings from one form to another, are the results of certain natural laws, the existence of which he discovers and proves by the history of the organic world.*

What a history of life was thus disclosed by Cuvier! Has any similar *fact* ever been

* See note III.

discovered? But he refused to see all this; denying the conclusions obviously resulting from his own researches. He took up a dislike to theories, seemingly because they were adopted and patronized by his academic rivals. Listen to his own remarks, "Theories I have sought: I have set up some myself, but I have not made them known, because I ascertained they were false, as are all those which have been published up to this day. I affirm still more; for I say, that, in the present state of science, it is impossible to discover any." The dogmatism and self-reliance brought out in such passages as the above, form the *forte* and *foible* of the race of men to which he belonged.

Thus he declared against theory, yet was himself the greatest of theorists; his great fact led to startling hypotheses, which he asserted to be facts. He maintained the *fixity* of *species* for ever and ever, grounding his assertion on the paltry pitiful records of humanity; records generally worthless, or so limited in time, as to be valueless for the settling of any great secondary law of Nature. The invention of six successive creations was at last forced upon him, chiefly through his English partizans; against his common sense, and in contradiction of his own writings.

All honour, notwithstanding, be to his great name; his dislike to transcendentalism was forced upon him. What passes for the views and theories of Cuvier, in England, do not belong to him. They emanate from a school, with whom truth in science is of no moment. They emanate from men who are not strictly scientific, but who, like Philo-Judæus, Derham, and Paley, look into works of science, not with any view to extract the truth therefrom, but to find happy applications in support of errors in human history, and a cosmogony to which antiquity has lent a sort of reverential awe.

Whilst Cuvier was still a very young man, and in Normandy, the thought occurred to him, that certain animals had been classed together by Linnè, in groups anything but natural. A deep consideration of their external characters would have told him as much, and must have enabled him to rectify most of the errors of the great formulist. He resorted to another method. He explored the interior, dissecting it with admirable precision and care. The result was, the discovery of the true nature of the animals we now call mollusca. A quarto volume on their anatomy, by Cuvier, excels perhaps all other monographs, saving always those of Hunter. As an observer, this work alone would have immortalized Cuvier with the scientific world; but the great world requires other discoveries than the descriptive anatomy of an oyster or cuttle-fish. These discoveries were soon to follow; Cuvier started then as a descriptive anatomist; from this view he never departed. It was his first and last labour, to examine into the anatomy of at least one adult individual of every species of animal; to describe it, and by doing so to show wherein it differed from all other species of animals. This was the anatomy of differences. The comparisons he himself made, were with a view to the discrimination of species, nothing more. They were comparative only in a certain sense. To these descriptions he tacked a specious general physiology, expounded in a clear masterly style, but leading to no great results. Fortunate it was for him and for science, that he persisted in such views and such researches. His progressive mind led him from the avertebrate animals, in which men take little or no interest to the vertebrate—that grand section of zoology to which man belongs; and having learned the art, clearly and distinctly, of discriminating one species of animal or one genus, at least, from another, by contrasting especially their skeletons with all others, he boldly launched upon that vast ocean of discovery, of which he at once took entire possession. On this sea, one bark only may be seen, even now; it was that which carried Cuvier.

Pirates, contrabandistas, appear from time to time on this great sea of discovery, chiefly English, who under pretence of pointing out a few barren rocks and sand-banks, which Cuvier had neglected to describe, or deemed unworthy of notice, conceal their scandalous calling; and how they live and fatten on the brains of genius! Their efforts are well understood. They tried the same game with his discovery of the application of descriptive anatomy to living zoology, to the result of which I again advert.

So long as Cuvier's reputation was confined to a few scientific men; so long as, in Britain for example, a knowledge of the

"Leçons d'Anatomie Comparée" was confined to some two or three in London, four or five in Edinburgh, six or eight in Dublin, the *compilateurs*, the race of the *flibustiers*, the men who work out the discoveries of others, in hopes of working out the discoverers themselves, took little or no notice of Cuvier or his labours. But when by the publication of the "Ossemens Fossiles," Cuvier's reputation became universal, then arose a mighty clatter amongst the dogged model men, the sticklers for the existing order of things; zoological collections (there were no museums) began to be re-arranged, brushed up, and set in a sort of order. Nothing of all this would have happened, but for Cuvier's discovery of the pristine world.

The history in fact of this great man is wrapped up in two lines; he first successfully applied descriptive anatomy to living zoology; by the same instrument of research he revealed the history of the pristine world: the vast mine of a world's reputation was touched; then started fresh and furious, the mechanical, hard-headed, utilitarian confederacy: "Follow Cuvier" was the cry, "and in the chase we may chance to outstrip him: and when dead our partizans will declare that his mantle has fallen on us."

THE LIVING ZOOLOGY.

All that tact and genius could do, aided by external characters alone, Linnè had effected. The great man whose life I sketch, saw this when still a youth. He was not vain enough to imagine that, as a formulist, he could excel Linnè, unless he brought into the field other modes of research. He addressed himself therefore to the internal structure, that structure which Linnè had neglected; in other words he dissected species, at first, perhaps even to the last, with a view to accomplish the great object of all his inquiries and researches, namely, the most natural classification of the living, and afterwards of the extinct animal kingdoms, according to their anatomical and *other* affinities. This was his great aim from first to last. The analysis of the fossil remains of extinct organic worlds was an accident, an episode in his methodical, mechanical, laborious inquiries, which took himself by surprise no less than it did the world. Let me consider first the application of his method, as it has been called, to zoology, which, in his view, means classification.

That scientific men should ever have endeavoured to classify animals merely through their external characters, must excite surprise only in the uninitiated. That they should neglect other sources of information, and especially the structure of the great internal organs, on which may seem to depend their intimate character, does at first sight appear strange.

But when we inquire calmly into the matter we find, first, that the assertions of Cuvier and his partisans on this point are quite exaggerated. Even the character of the digestive organs, so important no doubt in every zoological scale, may in general be made out by the teeth. To this, however, there are many exceptions. Thirdly, the character of the extremities gives the most extensive knowledge as to the nature of the animal and his place in zoology. Fourthly, the mode of reproduction, of lactation, degree of intelligence, food, habitat, temperature of the blood, all are circumstances which may perfectly be made out, independently of all dissection.

It requires no anatomical knowledge to see that a fish does not breathe like a mammal; whilst the absence of gills, and the necessity of breathing air, placed the meanest observer of Nature in a position to remove the whale and seal from that class of animals with which the unobserving mass of mankind, deciding merely from their habitat, had confounded and still confounds them. The merest peasant could surely determine the oviparous character of birds, fishes, and reptiles; the viviparous character of mammals. So far then, it is not true that naturalists, before Cuvier's time, had despised anatomy altogether. That their ignorance of this science was and is most profound, I admit, still they had done something in this way. It was reserved for Cuvier to show them that the day had come, when a precise, rigorous, and exact descriptive anatomy of species must be applied to Zoology. He may be considered then as the first who introduced it, systematically, into zoology.

In his "Règne Animale," distributed according to its organization," he even ventured to introduce the fossil mammoth, whose external characters, saving the teeth and a small portion of the skin, he had never seen; thus mingling the existing with the dead and living zoology, and unconsciously aiding in the putting forward an idea destined, shortly, deeply to affect if not destroy all his theories; mingling the organic worlds—the living and the dead. Even then had his grand instincts led him to suspect that if the two organic worlds were really not one, the dawn of the present extended much farther into the past than he had at first imagined.

That he greatly improved classification in zoology, none will deny. He all but created the avertebrate kingdom. Yet he was not happy in philosophical terms, which were generally invented by others. To Lamarck we owe the highly philosophic terms of Vertebrate and Avertebrate; to the same naturalist, the term Annelides; to De Blainville the formation of the class Amphibia.

Now all these improvements, whether effected by him, or others following in the same road, are due to the application of the exact descriptive anatomy of species to zoology. As Bichat was the discoverer of the importance of descriptive anatomy, as applied to man, and was therefore the discoverer of true descriptive anatomy, so Cuvier was the discoverer of exact descriptive anatomy applied to all other animals, inasmuch as he was the true discoverer of the importance of that branch of knowledge. What influence the writings of the former may have had over the latter I know not. The school of the Garden and the school of the faculty have ever entertained for each other a sort of jealousy; a kind of competition, I know not why or wherefore, exists between them even to this day. I remember a little *fracas* between Beclard and Geoffroy; it ended in words, and the surgeon withdrew from philosophy to his mechanical routine. In later times the illustrious Orfila made another attempt to introduce scientific anatomy into the faculty of medicine, and the Hall of Apollo contains some preparations of what is mistaken by many people for comparative anatomy! Vain attempt! The union of science with trade M. Orfila quickly found to be beyond his ability, great though it be.*

Cuvier laid it down as a principle, that the position of any animal in the animal kingdom could not be well determined, until its descrip-

* See note IV.

tive anatomy had been fully made out. This important deduction will probably never be overthrown. It is consistent with the common sense of mankind. That his "Anatomie Comparée," as he calls his description of the anatomy of species, began and ended the research is proved by this, that Meckel's great work, which appeared some years afterwards, received not the smallest notice from the world, scientific or fashionable. Men felt that the "Leçons d'Anatomie Comparée" of Cuvier had accomplished the aim of all: the work evolved the principles sought for, and henceforward men neither could nor would take any interest in the dissection of cats and rats, bats and weasels, flies and spiders. As is usual, our insular labourers, in what they are pleased to call the field of science, worked hard to share the honours of the field. Now it was a new species; then again a new variety; now it was a fossil reptile, rarer than the one described by Cuvier; or a bone described by Cuvier, was discovered stowed away in a cellar of the museum. They made large collections which they called museums; described and dissected a variety of animals;

gave themselves out as comparative anatomists; but all would not do; the comparative anatomy of Home is merely a subject for ridicule; and the osteological collection superadded of late years to the noble museum of Hunter, originates in the lamentable mistake of servile imitators, who know not the meaning of Mr. Hunter's grand idea. To follow *Couvier* (this is the way in which they mispronounce the immortal name), follow Cuvier was the cry; and try to surpass him! Collect more bones, grub together more fossil remains; describe them in detail even to nausea; and by running-a-muck, and getting partisans to make a great noise, the ignorant may mistake us for scientific men! With the ignorant they have partly succeeded; and men, to whose names there attaches not a single discovery in science, the announcement of a single new principle, have, on the shoulders of Hunter, and the author of the "Ossemens Fossiles," been foisted into temporary importance and notice. Posterity will put all right.

THE EXTINCT ZOOLOGY.

The actual value of a knowledge of extinct zoology to science is great, to the mass of mankind it is, in one sense, of little or no importance; but, in another, its consequences are incalculable, for by its means human reason burst the fetters of ages. Had Cuvier's labours been restricted merely to the production of an improved "Règne Animale," he must have, as in fact he now does, in that respect, figured as second to Linnè; it is the original work, the original thought, which alone is entitled to immortality. The efforts made by his partisans to invest him with the title of "Great Naturalist," are simply ridiculous.

It is quite otherwise with his grand application of descriptive anatomy to the fossil world. Before and during his time, and since, too,* naturalists, and geologists, and amateurs, were in the habit of judging of, and deciding on matters quite beyond the calibre of their minds. Spallanzani, the best of the class, spoke of waggon-loads of *human* bones which

* See note V.

might any day be seen in the south of Italy: of bones of men drowned at the Deluge, the only geological epoch which the vulgar can comprehend. Cuvier showed, that amongst all these waggon-loads of bones, there was not a single human bone. Naturalists and amateurs mistook the bones of elephants for giants; thus mingling up fable with truth. And Faujas St. Fond, a man who lived in Cuvier's time, asserted that the fossil saurians were simply crocodiles. Voltaire, an opponent of the Deluge theory, conjectured that if shells were found on the tops of mountains, they had been carried there by birds; and if a rhinoceros tooth had been picked up in a quarry, it had been left there by some naturalist! One man, George Cuvier, and one science, descriptive anatomy, put an end for ever to these insane and foolish ravings. How simple does his vast idea now appear! How natural! "If these fossil bones and shells, and plants, which you find in various strata of the earth's surface, really belong to species of animals and plants, which still live, then they ought at least to resemble them. The bones of an individual man, or ox, or sheep, or deer, resemble strictly those of all individuals belonging to the same species. The same law must apply of course to fossil bones." And then came out the astounding fact, that not one of these bones was identical with any species of animals now living, or that may have lived during the historic period. The origin of our historic period I assume to be limited to the monumental records of Egypt; of the world before that, nothing physical is known, saving through the lights of anatomy. Thus was opened up to man's view, the pristine world; not as we read of it in fabulous histories, the silly imaginings of foolish men, but as Nature made it.

We have now to view Cuvier as entering on a new career; from simple naturalist and anatomist, he had become geologist in a sense that never man was before. Historian of the catastrophes of the globe; author of a new cosmogony. Unwittingly, the man of facts was forced, also, to become the theorist. Human bones were not found fossilized. Then came theory first, namely, that man's origin or formation was quite recent. In theory second he advanced the principle of the fixity of species, founded on the fact, that during *the historic period*, animals had not changed their appearances,—forgetting that this historic period was but a drop in the great ocean of time; that no great geological epoch had occurred during that period; and, consequently, neither man nor animals had been violently dislocated from the aboriginal continents; ever been exposed to the only influences likely or competent to produce changes in form, amounting to a specific or permanent alteration. Lastly, a theory or two was forced on him by the theo-geological school of England, which were not his, and expressions which he never uttered were ascribed to him. It passes current, for example, in England, that he advocated the theory of successive miraculous creations of animals. This is a pure invention of the English geologists, invented by them to reconcile the conflicting facts of true geology, with their imaginary cosmogony and fabulous chronology. With the exception of his first paleontological essay, Cuvier constantly opposed the theory of successive creations. There cannot exist a doubt on this point, although the contrary opinion has, as is usual, become stereotyped in England;—in England where things are said never to change—not even errors. These are his words:—"nous ne croyons pas même à la possibilité d'une apparition successive des formes diverses." Thus the theory ascribed in England to Cuvier, this illustrious anatomist has declared not only to be false, but impossible.

Whilst Cuvier was thus applying with such success, the single method of the descriptive anatomy of the *adult animal* to zoology, and to the history of the globe, overturning all existing theories, cosmogonies, and histories, other minds were at work in Germany and in France. "The descriptive anatomy of the adult animal formed *selon le règle*," is not all; there is the anatomy of the embryo; of varieties or *lusus naturæ;* of monsters; of organs found in man and animals, evidently of no use to the individual. This Cuvier persisted in overlooking. His mind was filled with that idea, the most natural of all, namely, the persistence and fixity of the present order of things; an idea proved to be false, first by himself. But this, also, he would fain have overlooked. Of the new doctrines of transcendental anatomy, originating in these sources, he took but little notice at first, at times admitting them, at times rejecting them. At last the doctrine of unity of organization became too great to be overlooked; a struggle evidently approached between the parties. But it did not fairly come off until Geoffroy, the French advocate of the heterodoxy, had boldly advocated in the Academy, and in the presence of Cuvier, the theory of development, a necessary sequel of the grand law of unity, which teaches that "the animals to which belonged the fossil remains, so admirably described by Cuvier, are not specifically distinct from the living organic world, but simply the forefathers of the existing race of animals."

The history of the remarkable contest which followed, I shall give in my life of Geoffroy. Cuvier ought to have avoided its discussion. In fact, it did not in any way interfere with his great determinations—demonstrations, I ought to call them. But he thought so; and the world, which is worse, also thought so; and this forced on him the invidious task of assailing a theory, the correctness of which he had fully admitted in his youth.

The name of Cuvier will never pass away. Though following Bichat, he invented descriptive anatomy. He created a new science; a new philosophic instrument. By applying this instrument to zoology, which chiefly means classification, he extended, enlarged, and greatly perfected, method. He gave to zoology a sort of scientific character; at all events, he connected it with general science. By means of the same instrument of thought, descriptive anatomy, he discovered the history of the globe, in as far, at least, as life is concerned. But if life be coeval with the globe, which is my belief, Cuvier, then, gave us the history of the globe, by enabling us to read aright the vestiges of all former worlds in their organic remains. But this history is not philosophy, any more than any other history; the vestiges of extinct animals and plants are not vestiges of creation, but of that which has been created. The order in which they appeared, is even yet doubtful, and must long continue so. The vestiges of creation, which word can only mean the materials by which may be discovered those secondary laws, to which successive animal and vegetable forms owe their existence, were, in this sense, not discovered by Cuvier, but by his cotemporaries in Germany and in

France, and more especially by that man whose life and labours I mean next to sketch.

CONCLUSION.

As men decorate the idols they worship, so the admirers of great men are apt to bestow on them a reputation they did not merit, and qualities they never possessed. In the struggle to praise and to detract, truth is lost sight of; genius invested with practical ability, which it rarely is, or utterly despised. During a great portion of the life of Dalton, his name and reputation were held in the most sovereign contempt, even by his own townsmen; soon after his death, they were willing to pay him honours almost divine. This is human nature. It fared otherwise with Cuvier. During life the intellectual world held him in honour; detractors were forced to be silent, yet such existed. The chief or head (what a head!) of a bird-stuffing, shell-collecting establishment, which shall be nameless, used to be in the habit of remarking that natural history had been completely ruined by Cuvier, and the anatomical men. Let us hope, for the honour of the country, that few such persons exist.

Honoured during life by the thinking men of all countries, intellectual France, represented by its noble Academy, jealous of the reputation shed on their country by its adopted son, watched, perhaps still watches, and repels with energy, the slightest attempt to sully his great name. Soon after his death, his rival, as we must term him, Geoffroy, attempted to read a memoir to the Academy, glancing, though in a remote manner, at the possible overthrow of some of the great principles established by Cuvier. Though venerable by years, and by the unceasing scientific labours of half a century, the Academy refused to listen to Geoffroy. To assail the reputation of Cuvier, was to insult the dignity of France.

That his merits were great, nearly without a parallel, I, especially, will be the last to deny; I who have devoted forty years of my existence to similar pursuits; inquiries from which I have derived the most substantial pleasures of my life; but these merits are not such as have been represented by his biographer and friend, M. Flourens. On this point I think I have been already sufficiently explicit.

Comparative physiology, of which M. Cuvier

was, in some measure, the founder, is a science of but little practical utility, though important enough in the establishment of great generalizations. The facts of anatomy do not offer many fitted to form the basis of *a priori* reasoning. The external robe, and what may be seen externally, give indications as sure of the character of the animal, as any derivable from the anatomy of its interior. M. Flourens places then on a false basis, the claims of M. Cuvier to universal esteem. Cuvier restored natural history to science (in as far as it is science), from which it had been long separated. He applied himself to the discovery of the entire truth, as regards the present organic world; this led him to a clear apprehension of the unalterable and seemingly perpetually fixed characteristic differences which mark and distinguish, not more externally than internally, the various species, genera, and natural families or orders of animals, which people and decorate this world. Having done so, a bright flash of genius completed the rest. It led him to apply the law of co-relations of structure of functions, and of zoological positions, to the fossil bones, which by all, or nearly all before him, had been taken for the remains of animals drowned at the Deluge. He applied also with great success, a law, which, in whatever light it be now regarded by transcendentalists, is still practically true, namely, the fixity of species. The results have been already placed before the reader. Henceforward, Cuvier, the naturalist, which properly speaking he never was, although it pleases many to call him so, henceforward Cuvier, the anatomist, took his place in the page of history, side by side with Aristotle, Leibnitz, Newton, Galileo, lovers of the perfect, of the true; names which neither time nor circumstances can erase from the memories of men.

SECTION II.

Etienne Geoffroy (St. Hilaire),

HIS LIFE AND LABOURS.

COEVAL and cotemporary with George Cuvier was Etienne Geoffroy, who, assuming early in life the patronymic name of St. Hilaire (where he was born) came at last to be called St. Hilaire by English naturalists, alike incapable of comprehending his name, his genius, his position in science. Geoffroy's career was singular, most singular; it was also somewhat unfortunate. Endowed with the highest genius, he was disliked by all the mechanical world around him, within and without the walls of the Academy; in the Jardin des Plantes he stood his ground with difficulty.

During the life of Cuvier, and even afterwards, the Academy still remaining in the hands of the partisans of the Cuvierian School, Geoffroy's chance of success was small. How could it be otherwise with the man of genius, the man in advance of his age, the abstract reasoner, the observer only of abstractions, the man who, with Autenrieth and Goethe, saw in the structures of man and animals, not what, to the ordinary minds of men, they *seem to be*, and as they must so remain to the end of time, but their signification; what they really mean, what in fact they are; who established on a basis, not again to be shaken, the doctrine of analogues and homologues, that is, unity of organization, and unconsciously first gave to the modification or extension of that doctrine called the theory of analogues—a theory which the hyper-transcendant views of Oken and Spix had injured—a secure foundation in positive and well observed facts. How could such a man succeed as the opponent of the perpetual secretary, the illustrious Cuvier, Baron of the Empire, Member of the Council of State, President of the Board of Education! What chance had such a man against him who formed in himself an era? The very students declined attending his lectures; he was looked on as a well meaning *savant* somewhat deranged. And when to this was added the consideration, that he, Geoffroy, ignorant of anatomy, comparative and human, was opposing on *anatomical grounds* the person whom the voice of Europe, right or wrong, had placed at the head of the anatomists of his day, it must be evident on which side, for a time at least, the victory was sure to incline.

And now, Cuvier being but, as it were, dead, his era is gone; it is in vain that the mechanical, plodding, descriptive anatomists, whom he found, introduced to the Academy, and left in his place, can ultimately maintain it; it is in vain that M. Flourens, representing these mechanical utilitarian minds, asks for a compromise, begs of you to see in Cuvier's views one great truth at least, and one which need not exclude another. This were well if the question merely involved two facts; but it is not so; the question is with a theory, the greatest ever offered to the consideration of man—a theory which Cuvier and his followers rejected and reject; a theory which is either entirely true or entirely false, which neither admits nor requires any compromise, any support; a theory which says to the philosopher, "the classification of animals and vegetables is not the aim and end of natural history—classification is not the philosophy of zoology;" leave such mechanical utilitarian minds, with their contrivances and adaptations of providential resources, their perpetual chorus of "wise provisions," their specialisms and individualisms, for in this they end at last; their outrageous anthropomorphologicalisms of the first cause, their denial of secondary causes, their ten creations and fifty submersions: leave them, says this great theorist, and with me endeavour to see in Nature one system; and connecting man with the organic world, the existing organic world with the past and with the planetary system, that past system with the universal, endeavour thus to discover in these relations, the great problem of Man's Creation.

Prior to the appearance of this remarkable man, philosophic minds of various ages and various pursuits had announced the bold theory, "that all animals are formed upon one plan." Leibnitz, the great rival of Newton, entertained this opinion; so also did Newton. Pascal threw out in like manner this grand conjecture, for until the transcendental in anatomy arose, such it merely was. Bacon had recommended experiments to be made in order to discover the causes of forms. The mere observance and classification of forms of life did not satisfy these great minds, these lights of the earth. They desired to know whence and how originate the various forms which life assumes on this globe. What cause or causes—physical causes, they thought of none else, nor can any other even be imagined—give rise to the indefinite, if not infinite variety of forms which have decorated or still decorate the earth. This, the greatest of all philosophic questions ever proposed, next to the origin of the globe itself, was thus formuled by men, not in themselves naturalists or anatomists, but who possessed a genius equal to observe all material phenomena. They considered this question as it really is, a natural, a physical question, a question of secondary laws.* But they only conjectured; theirs was an hypothesis merely. Buffon attempted its solution, but still as a mere theorist. Last came Goethe, Oken, Autenrieth, Geoffroy; they attempted the demonstration of the causes of forms, and if they failed in this, as many still think they did, they at least proved that the living

* See note I.

organic world and the past have been formed upon one great plan, one scheme of Nature, the basis of which is the unity of structure, unity of organization. The discussion of this great problem gave rise to others more or less directly involved in its solution. A minute descriptive anatomy, a correct descriptive anatomy of the various species of non-existing animals led Cuvier, as we have seen, to correct, first, what was defective in the classification of Linnè; second, to detect the real nature of the fossil remains of animals which had once lived on the earth, and to show that they differed specifically and generically (admitting species to have a distinct existence in Nature's plan), from those now existing—the greatest discovery ever made in science; and, thirdly, he improved by this more correct anatomy, comparative physiology, a matter of little or no consequence to man in a utilitarian point of view, but of infinite consequence when viewed in relation to science.

If the enlightenment of the human mind be the highest possible aim of intellect, if the discovery of the truth (and such a truth was wrapt up in the history of the "Ossemens Fossiles"), if the discovery of truth be the greatest achievement of human intellect, then does the claim of Cuvier to the universal thanks of mankind stand undisputed.

But a more minute anatomy, known only to the few, had arisen even before his era commenced; inquisitive original minds had adopted a nicer observation of structure, descriptive also but no longer applied to the regularly formed adult animal, the great object of the Cuvierian researches. The anatomy I speak of seems to have originated in Germany; not North Germany, the land of schnaaps, and insolence, and dolt stupidity, the land of the Pruss; but in south and middle Germany. Autenrieth of Tubingen seems to have been amongst the first who clearly understood the principle; but I am not now tracing the history of the discovery of transcendental anatomy, but its application partly in Germany, chiefly in France, to the philosophy of the organic world.

The anatomy I am now about to speak of, treated of all those mysterious structures which the other, essentially the special anatomy of the adult, had neglected, or purposely, or unintentionally overlooked. The anatomy of Cuvier and of Bichat, as we have already shown, had a reference, exclusively, to the fully developed adult—the animal specialized, the type of his species, his genus, his race.

The anatomy I am now about to speak of, does not despise or neglect this specialism,—this individualism. But it sets it down merely for so much as it is worth. It admits it to be the basis of all true zoological knowledge, but it views it as part and parcel only of anatomy, not the whole. Accordingly, Autenrieth, and his German coadjutors, Goethe, Oken, followed by Geoffroy and his school, proceeding on the great ideas of Aristotle, Leibnitz, Pascal, Harvey, Newton, maintained that, besides the anatomy of the adult and specialized individual, there were other structures to be examined, other facts to be discovered and explained, other laws to be determined. Select, they observed, any system of organs belonging to man himself, —the osseous in preference, as being easiest handled and observed, most enduring, and, perhaps, the most characteristic,—select this set of organs as a subject of inquiry, and take an adult regularly formed man for your type, and you will find, first, that all are not precisely alike, that they present varieties, the causes of which—your silly, unintelligible name of *lusus naturæ* being rejected by all thinking men—remain to be discovered.

Amongst a thousand such varieties in adult human structure, which not unfrequently present themselves, constantly overlooked by the mere anatomist of adult forms, we may here notice the webbing of the fingers and toes; the overlapping of the bones of the cranium; the hare lip; the apertures seen occasionally in the lower part of the neck of man; the absence of arms and legs, the feet and hands being present; the semi-bent position of the elbows and knees; the elevated calf of the leg; the projecting heel; the lengthened great toe; disproportions between the trunk and limbs; a stomach composed of two compartments instead of one; hypospadias, &c.

These mysterious varieties taught no lessons to our predecessors, wrapt up in practical utilitarianism; mythical cosmogonies; a contempt for truth; fatalists; admirers and believers in "the best of all possible worlds." *Lusus naturæ* was still the phrase: Bismillah! ejaculated the orientalist; wonderful are all thy works! responded the western fanatic. But the torch of science, quenched deeply in the mire of horrid ignorance, and, what is still worse, a learned pedantry, since the time of Aristotle, still glimmered—was not extinguished: It burst forth in the German school. Amongst the earliest to announce the new generalization was Autenrieth.* "These varieties in structure, which you decline inquiring into, are not hap-hazard formations; they are the remains of structures common to all embryos; they indicate the transitions through which man, and all other animals, are passing from their embryonic condition to the adult." Should anything interfere with this transition, the embryonic, or infantile, or juvenile forms, persist to the adult condition. This constitutes what you call a variety, and which you mistake for something new; a phenomenon over which other laws preside. But it is not so. The laws of deformation are as regular as the laws of formation; the varieties you observe are simply caused by "arrests of development." These arrests of development show you what the being once was; they prove that variety in individual structure, is but a return to unity of or-

* The work has been ascribed to Harvey. I doubt the truth of this conjecture.

ganization; the unity of type, if not of germs, with which Nature starts in the formation of all that lives. I need scarcely add, that the exact descriptive anatomy of the embryo demonstrated the correctness of this theory. Thus was the great law first announced, which was to revolutionize human knowledge. But it did not and could not stop here; a second law was speedily discovered.

The same illustrious observers were not slow to remark, that the adult structure of the lower animals strongly resemble these embryonic forms; or, rather, that in his embryonic forms man strongly resembles the adult structures of the lower animals. A second great law was thus discovered, namely, that the embryonic forms of man shadow forth the range of the animal kingdom as it now exists. It has been objected to this view,* that an embryo man is not a worm, nor a mollusc, nor a fish, in succession; that he merely resembles these forms of life. Be it so. No one ever said that he was any of these animals at any time. What was said, is, that he resembles them in his organization; that he

* Flourens.

has their forms; that his organs have not human forms, but bestial or animal; that as an embryo he has gills and lungs; that as an embryo his brain has not a human but animal form; that the digestive organs resemble strictly those of animals much lower in the scale; that his heart is not double, as in the adult, but single and piscine. To deny these *facts*, you must be prepared to deny the observations of the greatest anatomists of modern times.

A third law necessarily followed this, which, if not the first to announce, I was, at least, the first who attempted its demonstration. The law of unity of organization reposed mainly on the fact, that rudimentary organs existed in many animals, distinctly proving a unity of plan—even of germs. But these are often wanting. Organs, also, exist in all animals, whose uses are quite unknown.

First.—In man, philosophic anatomy reckons three, four, or five, cranial vertebræ; thirty-three as belonging to the trunk or torso; but in many animals there are many more, amounting sometimes to more than two hundred.

Secondly.—Varieties in the structure of the

adult man appear, strictly shadowing forth constant forms in the lower animals, but which do not necessarily form a part of the usual embryonic structures. Such, for example, as the extension of the supra-condyloid process (a mere rudiment in most human arms) to an osseous projection, almost completing a supra-condyloid aperture, through which passed in the human arm, observed by me, the main artery of the arm and the great nerve. Now this is the *regular structure* in all the feline tribe, the lion, tiger, panther, &c., and perhaps in some others. But as no such structure is constantly observable in the human embryo, it cannot be that such a variety in man is an arrest of a development, since such a structure, as a constant law, does not exist.

Thirdly.—Organs exist whose uses are wholly unknown.

To meet these difficulties, I ventured to offer a third great law, or generalization, namely, that a great plan or scheme of Nature exists, agreeably to which all organic forms are moulded. That the precise type, forming the basis of this great law, can never be fully discovered for this reason, that it embraces all life from the beginning to the end of time. When the laws of specialization are interfered with, or, as it is sometimes expressed, the laws of development, the variety produced may either represent a mere arrest of development, or it may represent some other form, or possible existence embraced in Nature's scheme. This form may either represent the sub-type, when perfect, of a now existing animal (as in the instance recorded above), or it may resemble an animal form now extinct, or one not yet called into existence.* "That (type) which you seek for is nowhere to be found, and yet is everywhere." †

Fourthly.—The inquiry did not stop even here. Monstrous productions, as they are called, were appealed to; the anatomist of regular adult forms took no notice of these: but Geoffroy and the German school did.

Fifthly.—The law discovered by Cuvier, that the ossemens fossiles, or organic remains of a former world, are specifically and even generically distinct from the existing races of animals; the *fixity of species*, limited, however, by Cuvier, to the historic period; and the

* 1827, Lectures and Memoirs. † Plautus.

successive generations of man and animals came to be openly disputed in that Academy, where Cuvier reigned triumphant; and in the face of Europe. This was the language of the transcendentalists:—

"The differences which you consider as specific and generic, may be merely anatomical differences, produced by a succession of ages, but not by new generations in the sense you view them. This problem, which your partisans at least assert to be a physical one, may after all be but an historical phenomenon." Cuvier was but dead when De Blainville proved that it was so.

The life of the man who was so bold as to attempt this, against such difficulties, deserves a place in the biographies of illustrious men, even admitting that some of his theories are still open to objections. I shall sketch his personal history;—the episode of his journey to Egypt, and to the Iberian peninsula, concluding with a few applications of the theory of unity of organization to the organic world; past, present, and to come. But first to unfold the principles of this great theory, I must again advert, however briefly, to the history of zoology, prior to the period when Cuvier and his illustrious rival and coadjutor (for Geoffroy contributed also his share in placing zoology on a proper basis), by applying to it the principle of sound descriptive anatomy, restored zoology once more to science and to philosophy, from which it had been separated for more than two thousand years.

ZOOLOGICAL CLASSIFICATION.

There are entire ages which form no epoch, no era; the leaders in such ages I need not name. Time sweeps them from the recollections of men. I cannot find the period when there was an epoch of natural science in Britain. The "Three Hundred Animals" was a classical book in my younger days. I believe it is so yet. The actual condition of physical geography in Britain, during the early part of my life, surpasses all belief. The gulf which separates the man of science from the man of letters, was not, and is scarcely yet understood. Pliny and Aristotle were viewed as belonging to the same class of minds; Goldsmith and Smellie wrote about the philosophy of natural history! and theologians, and even handicraftsmen, whose education consisted merely in reading, writing, and casting up accounts, were considered as high authorities in natural history. This disgraceful period has not altogether ceased in Britain, and I can even imagine circumstances in which it might return. In the absence of all scientific knowledge, final causes were on every difficulty appealed to, and at one time England was a kind of modern Prussia; the minds of men had no alternative; assume the manners, the livery of the day, or perish.

The knowledge which has to man no practical bearing; which is neither directly useful to him in a utilitarian point of view, nor illustrative of science, as part and parcel of the great principles of the highest form of human civilization, never has called and never can call forth from the mass of mankind, any general sympathy. Thus it was with what used to be called, and in some ancient non-progressive institutions, is still called, natural history, meaning the names and classification of plants and animals, with definitions, interminable disputes about their identity or differences; a displacement or an adjustment of species of animals, concerning which mankind not only takes no interest, but wonders perhaps at their existence.

That great minds should despise such frivolities, such twaddle, such puerilities, in the hands of mediocrity, was not to be wondered at. But when this science of observation as it is called, passed from Aristotle into the hands of Cuvier, of Geoffroy, of Oken, and Goethe, the minds of all thinking men over the earth were irresistibly called on to watch the results. Of these I have already described one: the discovery of the true history of the earth by Cuvier, the founder and discoverer of stratigraphic geology; the next I have still to speak of; the philosophy of that history of the organic world in time and in space.

To Aristotle is due the merit of having attempted the solution of both problems; the classification of the living organic world, on a natural plan, based to a great extent on structure; second, of announcing the great theoretical idea of unity of structure. But true descriptive anatomy was not understood in his days; nor long after. Vessalius and Della Torre failed in putting it on a proper basis; the labours of the illustrious Dutch and German schools were confined to monographs; nor was it until Bichat and Cuvier appeared, that the part which descriptive anatomy was to play in this history of classification and of philosophy, came fully to be understood. From Aristotle to Linnè, there existed nothing, because there were no original observers. There was no progress, no advance of the human mind in respect of these sciences. In the second period appeared Linnè; this was in the eighteenth century. Thus, for nearly two thousand years the human mind had stood still in respect of natural science.

The name of Carl Linnè is immortal. It will be for ever fresh and young in men's minds. Like the antique statue, it never becomes antiquated or old fashioned, as happens with most mediæval and modern ideas. Compare the last poetasters of modern England with Horace and Burns. It is ever the same with genius.

The tact of Linnè in zoology was admirable. Cuvier merely followed him, extended his views, and corrected his arrangements, faulty in consequence of want of data, want of information, want of assistance. Haller, the envious, captious, invidious Haller, grudged him his reputation, and would have crushed him if he could. If Linnè invented the artificial method in botany, he also invented the natural method in zoology. About the same period appeared Buffon. The style of this profound thinker's works was so admirable, so dazzling, that people forgot in his literary merit that in some respects he was also a scientific man. I say in some respects, for as neither Linnè nor he was acquainted with descriptive anatomy, so neither was it possible for either to follow out exact science. But with a lofty genius, leading to abstractions he went further than Linnè, and speedily attracted the notice of the Sorbonne; with that redoubtable foe "he measured swords and parted." It was a pretty little quarrel while it lasted, but led to no results. And now in the train of these great men followed an illustrious school, Pallas, Blumenbach, Trembley, Camper, Lacepede, Meckel, Rudolphi, Latreille, Lamarck,—last and greatest, Cuvier and Geoffroy, destined to overshadow them all, and to form an epoch or era in science.

To Belon is ascribed the first attempt at a *demonstration* of the fact that man and animals are formed upon one plan. He placed the skeleton of the bird beside that of man, and endeavoured to show in what they *resembled* and in what they *differed.* Unity of organization was undoubtedly the aim of the demonstration, imperfect as it was. He saw the analogous, and the homologous; these are not two theories but one; the wearisome laboured verbiage lately got up in England, to point out some trivial differences in their application, aim simply at so mystifying as to conceal the names of their real discoverers. The scheme is coarse and hackneyed, despised even by their partisans.

John Ray next appears in England, and begins to classify more methodically, and Perault and Duverney founded zoological anatomy in France. Leuwenhoek, Swammerdaen, and others greatly advanced this kind of knowledge, which, however, was far from philosophical, but in the right direction. Some men in England attempted the same path, but they were insulted, abused, and slandered by the corporations; in Italy the minute researches and beautiful discoveries of Malpighi drew upon him a storm and a combination of "the practical men of his day," the men who "holding office, dislike change," which it required the authority of the Grand-duke to quell. A solution of the great philosophic question wrapt up in the life and labours of Geoffroy, had been attempted before his day by Buffon, Lamarck, and others. Their views were wholly theoretical. Lamarck's idea was that organization was the result of function, and not function the necessary result of form; that an animal was aquatic, not by the nature of its organization but became so, acquiring a fitting organization by its being forced to live in water. This view was wholly theoretical and met with no respect. Isidore Geoffroy, son of the illustrious Etienne, asserts that the real merit of the discoveries in fossil geology belong, not to Cuvier, but to Buffon. Certain it is that Buffon threw out the idea, but Camper, and Pallas, and Blumenbach had done the same. In the hands of Cuvier, as we have already explained, it became a demonstration no more to be shaken than the Principia of Newton.

I have thus brought down the history of classification and of the philosophy of zoology to the period when Geoffroy appeared. In Germany men's minds were agitated, and foresaw that a step forward was to be made in human knowledge. In France there was less enthusiasm; two most exact anatomists and physiologists held supreme rule over the scientific minds of the day; including Geoffroy; Cuvier, and De Blainville. I first met the illustrious trio in 1820–21, and became speedily aware that a crisis approached. But circumstances delayed it long, and it was nearly ten years before the final struggle took place between the descriptive anatomist of adult forms and the transcendentalist. By this time the works of Spix, Oken, Meckel, Geoffroy, and a host of others, and in Britain my own lectures, from 1824 to the period alluded to, had made all reflecting men acquainted with the fact that a new philosophy had appeared; that, astounding as had been the discoveries of Cuvier, upsetting all human history, they were likely to be eclipsed by another, based on the descriptive anatomy of animal structure philosophically viewed; of *lusus naturæ*, and of the human embryo.

Before I enter on the history of this remarkable era, it seems right that I should sketch briefly the life or personal history of the man who proved so instrumental in establishing on a firm basis, a theory broached by Aristotle, Pascal, Newton, Leibnitz; and secondly, the origin and progress of that Institution without the aid of which it seems impossible that such results for science could ever have been attained.

THE GARDEN OF PLANTS.

What is Natural History? What is its aim or object? Why should men study natural history in any or all its departments? What is Comparative Anatomy, its aim and utility? What is Transcendental Anatomy, and who discovered it? Finally, what has the study of these seemingly unimportant sciences to do with the history of man, of the living world, of the globe? How can they illustrate the history of the universe? Questions like these are seldom put by the young. Still seldomer by the old. The first is starting on a career, the aim of which is seldom explained to him, or if attempted to be explained, the explanation is for certain false. The latter has run his course, and in him, in all probability, the desire to know more has become extinct. The chances are ten thousand to one that his mind, as regards science, is not progressive, and therefore he halts and ponders. If superficially educated at first, he declares all science, literature, and art, to be mere folly and vexation of spirit; its pursuit especially, which with him has now lost its charms. Not so with the few of progressive minds; they pursue inquiry to the last; with them it is sufficient that the object be unknown, not well understood. Science they pursue for the sake of science, each after his own way and race.

I write this parallel biography with a desire to answer some of the questions I have placed at the head of this short introduction. In the course of it some truths, not generally known, will be brought out. Two of the four great men, whose lives and labours I purpose sketching and contrasting, were my senior contemporaries; they were also my personal friends. Their inmost thoughts on the nature of science and scientific research were known to me. Lamenting their differences and disputes, holding both in equal esteem and deep regard; aware that although not great or first-rate men in themselves, they yet constitute an era never to be effaced from human history, I have thought it my part to write a parallel biography, that the position each held in that scientific era may be rightly understood. It is due perhaps to the insular position of Britain that the era I speak of has been so singularly misrepresented in this country, partly no doubt to prejudice, partly to the material utilitarian formulas with which the national character clothes all exotic works. But be this as it may, certain I am that not merely has this era been misrepresented and misunderstood, but efforts are continually being made to stereotype these errors, to clothe them with British collegiate formulas, and to give to these formulas the perpetuity which ought alone to belong to truth.

The works of the heathen Aristotle, who believed in nothing but what was material, had been early formuled by the Catholic church and proclaimed as eternal truths. In England, under the hands of the Anglican church, they assumed another form. Aristotle and orthodoxy was still the war-cry of the schools. Now it is Cuvier. How this great man was all but forced to say what he never meant, and then was represented as having said so, of his own accord, is a part, and not the least curious, of this biography. But all this I have already explained.

In the winter of 1845–46, business of a scientific nature called me again to Paris. Requested by a commercial firm to proceed to the centre of taste and civilization, Paris, to confer with the discoverers of an ingenious process of great beauty and utility to medicine, I was led, by the very nature of my engagement, to visit, however briefly, that scene of the labours of my esteemed friends, whose names ornament the title pages of this little work, of whom two were already numbered with the dead; the third, De Blanville, alas! too soon to follow. Of my three friends, there remained but De Blainville. Him I sought at the Jardin des Plantes, occupying, I think, the house of Lamarck. Geoffroy had ceased to be, and the illustrious, and surely we may say, the immortal, Cuvier, had quitted for ever the scene of his labours and his glories. These men I had seen and been intimately acquainted with. The views they took of science and scientific men, were perfectly well known to me. So were their labours.

I have said, that my object in visiting the French capital was business and not amusement. This business led me to the Jardin des Plantes; to the School of Medicine; to the hospitals; to the Museums. A part of what I saw in the few days which this visit occupied, I had better state here; it will serve as an introduction, not inappropriate, to all my future remarks.

On the southern bank of the Seine stands the Academy of Science, Literature, and Art, usually called the Palace of the Institute. Behind, and around it, is the Pays Latin, that is to say, the learned quarter of Paris; and here, at various distances, will be found the School of Medicine, the Garden of Plants, many hospitals, and the University itself. The Sorbonne, with its dreaded recollections, stands also here; great national schools; colleges of high instruction; the Polytechnic School is also, I think, on the southern bank of the Seine; the *habitat*, indeed, of literature, science, and even art; contrasting strongly with that turbulent northern shore, the seat of fashion and debauchery; of political struggles; of regal power (as it then was); of wealth and pleasure; the Salle Valentino and the Madeleine; the Boulevards, and the *bœuf gras*.

It was into this quarter, I mean the learned side of the river, that business and inclination now led me. I had paid it a hasty visit, no doubt, in 1825–26, but it might truly be said that since 1820–21, I had not visited the Garden of Plants nor museums. Thus a quarter of a century and more had elapsed. I will compare, I said with myself, as I hastened thither, what now is with what was, and with what may never be again. I will note the changes, if any; the improvements, should such exist. In Paris, as is said, fashion is ever changing, and there are fashions in science as well as in dress. What is the present "mode" of science in Paris? This was my first thought.

On entering the Garden of Plants, the Salon Anatomique drew my first thoughts towards it; the creation of the great Cuvier, the scene of his labours and discoveries. In the outer court, in the position and condition in which I last saw it, but now covered with green mould, lichens, and those cryptogamous plants, which as surely mark decay and ruin to tower and castle, as grey hairs betoken the approaching fate of man, stood that mutilated skeleton of the Cachalot, for which

Cuvier had paid, imperfect as it was, one hundred guineas. And now entering the various apartments, containing the osteological and other sections of organic life, which formed the basis of all his views, I found them gloomy, deserted, mouldering, decayed. Additions there were none,—none, at least, that I could observe — and I involuntarily exclaimed to my most esteemed friends and companions, MM. Thibert and Percy, "I now see that Cuvier is really dead."

In the Museum of Natural History, fanaticism and prudery had been at work. The collection was in all respects neglected. The nude marble statue of Venus had been removed from the centre of the museum, and thrust into a low, damp, underground cellar-like apartment, amongst the dolphins. Dust prevailed everywhere. It was now a Celtic museum; Cuvier, its German founder, was gone, and with him the wish or desire to maintain that great work, on which reposed the scientific zoological era, he had created—of which he was, indeed, the originator, the centre. Thus great and original-minded men lay the foundation of vast intellectual monuments, which their successors are destined, as in the instance of Mr. Hunter's successors, to misapprehend, to neglect or even to destroy: at times, by misunderstanding the object of their founders; at times through envy; occasionally by neglect.

Before Cuvier founded his comparative anatomical museum in Paris, John Hunter, the most remarkable man of his time, had founded a physiological museum, a more remarkable work than the wall of China. His very object has been misunderstood by his followers, for he had no successors properly so-called. They have turned his grand physiological museum, illustrating by anatomy the laws of life, into a Dutch collection. He dissected and prepared structures to display the various forms of the organs by which in the living world the same function is performed. Cuvier's view was different, although it also embraced at first, unconsciously no doubt to himself, the physiological view of Mr. Hunter. He dissected and prepared structures; first, to show the different forms which the same organs display in different classes of animals; secondly, the exact descriptive osteology of all the living and now-existing vertebral world; without such knowledge, the precise place of each animal in the classification

E 5

catalogue of living beings, could not be determined; without such knowledge, the fossil world must have remained for ever unknown. This led to the formation of that matchless osteological collection, matchless not for its extent, but as being the first ever formed; strictly anatomical, it was mainly osteological, though having but slight reference to the physiology of locomotion; it was formed to determine the species of living animals, to contrast them with each other, and with the fossil dead; to enable the observer to trace the descriptive anatomy of at least the passive organs of motion, the only durable remains of mortality, and through them to guess at still higher views.

When my esteemed friend De Blainville said, that the anatomical descriptions of Cuvier were not comparative anatomy, he expressed an opinion strictly true in one sense, but not in another, and he mistook or undervalued the objects of Cuvier's researches. He himself taught comparative physiology, and taught it in a way in which he never had an equal. But Cuvier's works were at once descriptive and comparative; mechanical and philosophical; tedious, unreadable, minute.

What results have not this museum, formed by Cuvier, led to!

The principle of life, aimed at by Hunter, remains still a problem, not yet solved; but Cuvier, the mechanical, plodding Cuvier, has solved the greatest problem ever proposed to mankind (saving one), namely, the chronology of the globe we inhabit. He placed stratigraphical geology on a sure and solid basis, which before his time was a dream, a fable.

Had Newton not lived (Newton and Galileo), to what a deplorable condition might not the human mind have been reduced! A wily, heartless priesthood are now engaged in forming in a neighbouring country a university (!) in which the doctrines of Newton are to be refuted! It is to be shown that the earth is very large, and the sun very small! that the earth moves not, whilst around it revolves the universe! I stated publicly, years ago, that it would come to this; that the race amongst whom this has happened would never alter. But make of the globe and the sun what you like, you cannot deny the fossil remains discovered by Cuvier; the "Ossemens Fossiles" will prove too hard for you. There stands the barrier you cannot shake, against which the dark and hideous ocean of superstition will roll its surge in vain.

But I must return to the Garden of Plants, and to Paris, and to the history of zoology, that I may bring this episode to a close.

Were this great man's views understood, even by his cotemporaries or immediate followers? I doubt it. Apart from the zoological and anatomical halls, and at a considerable distance, but within the precincts of the Garden, I discovered a large and handsome salon, which to me was new. Here then at least was what seemed progress. But it was not so in reality. It turned out, on inspection, to be the Hall of Geology and Mineralogy, once more separated and detached from that museum, the comparative anatomical, from which it sprang, and to which it owed its being. In this hall amidst a profusion of the most beautiful mineralogical specimens, there stood the marble statue of Cuvier; the globe of the earth at his side; a finger points to the unknown regions of Africa, as if he had been a traveller—a Mungo Park—a Humboldt. Disjunction of men and things! dislocation of ideas! what had Cuvier the anatomist to do with mineralogy and geology? He was neither a mineralogist nor a geologist, in the strict sense of the terms; he was an anatomist of living and dead organic forms; the former he dissected with the scalpel, the latter with the chisel; he was a descriptive anatomist and demonstrator, and his anatomical demonstrations upset all existing theories of the earth, all chronologies; the history of the organic and inorganic worlds, as given us by historians, he showed to be a fable, unworthy the notice of scientific men.

But although I saw that Cuvier was really dead, his era determined and closed up, science had not ceased to exist in these gardens; De Blainville still lived and laboured. The history of what I saw in his laboratories I shall give in the body of this memoir. The amiable and excellent Valenciennes devoted his attention to the completion of Cuvier's laborious work on Fishes; the cedar planted by Jussieu, and the poor collection of living animals recalled associations of other times; of Buffon and Daubenton, Lacepede, and Latreille, whom I had seen. It was winter, and though Parisian, the Gardens were deserted, never more, I fear, to be sought for by strangers from all countries. Natural history was no longer fashionable, anatomy not thought of, geology had assumed another form, and decoration and art occupied all minds. But it was but seeming and not real, as the terrible events which so quickly followed the period of which I speak, soon set forth.* It was a calm before a hurricane, which in a few stormy days swept off a monarchy of its own adoption; returning once more to the worship of that name on which repose the glories of France; the mighty deeds she has achieved, and to whom we partly owe the era of which I now speak.

LIFE.

Etienne Geoffroy (St. Hilaire) was born at Etampes, on the 15th of April 1772, of a family of small means, originally of Troyes,—Troyes, infamous in the history of Napoleon the Great. Educated at Etampes, where there was a sort of college, he was intended for the church; with this view a bursary at the college of Navarre was bestowed on him, and a living offered; but Brisson's lectures

* The expulsion of the Orleans family from the throne and soil of France.

proving much more attractive than theology, young Geoffroy abandoned the church, and, to please his father, became a student of law. Remaining in Paris he completed his philosophy, as the wordy metaphysics even yet taught in universities are still amusingly called, in 1790, and entered himself a student at the Jardin des Plantes and College of France. He became a resident in the college of the Cardinal Lemoine, and was made Bachelor in Law also in 1790. But here his law studies ended. There remained to him, as some may suppose, but one change more, that is, to medicine, and to this he betook himself, becoming a student of medicine. But he changed once more; rejecting theology, law, physic, as conjectural, he took to science, that is, to the pursuit of truth, ever afterwards remaining a strictly scientific man.

It was at the college of Cardinal Lemoine he first met with Haüy, the illustrious crystallographer. They became intimate friends. In 1792 ('92 the terrible) Geoffroy was only twenty years of age. At the Garden he met Fourcroy, Daubenton, and other purely scientific men; and "the art called conjectural," *par excellence*, lost all charms with him for ever. He abandoned his medical studies for pure science.

Through the interest of Daubenton he gets a footing into that Garden where he was to pass so many happy and so many turbulent days. The tenth of August arrives; some of his best friends are priests *non assermenties*. Haüy, his patron, was arrested first; by Geoffroy's exertions Daubenton and the Academy, roused to a sense of Haüy's danger, demand the liberation of the accused. He seems to have saved the life of Haüy: his son, Isidore Geoffroy, asserts this, in the most positive manner. His personal, or, at least, his moral courage, must have been great, for at midnight he placed a ladder against the walls of the dreaded prison, after he had failed in relieving the other priests, by dressing himself as a commissary of police, and requiring their removal. But they one and all declined to leave the prison, unless all were liberated. Thus foiled, young Etienne, in the dead of night, placed a ladder against the prison wall. By this ladder, on the top of which he stood, on the 2nd of September, whilst the tocsin was sounding the live-long night, he rescued these unhappy men. The priests had refused to leave the prison with him during the preceding day, when he entered under the assumed garb of a commissary of police, and next day he witnessed the dreadful massacre, and an aged priest thrown from a window of the prison. It was on the following night that he placed the ladder against the wall of the prison, and scaled its summit: for eight hours no one appeared; at last twelve victims escaped; but the daylight came on, and a shot fired at Geoffroy from the garden of the prison, apprised him of the danger of his position. In two days afterwards he fled to Etampes. A furious fever was the result of those efforts, from which, however, he speedily recovered. Some months afterwards we find him once more in Paris, in the beginning of the winter of 1792. Daubenton, now his personal friend, assists him in his progress in life. Lacepede being forced to leave Paris by the progress of the revolution, St. Pierre, at that time the superintendant of the Garden of Plants, names Geoffroy to the humble office of keeper and sub-demonstrator at the Garden, situations held by Lacepede. Daubenton was keeper and demonstrator. He became thus Daubenton's assistant at a moment when ruin seemed to threaten not merely the Garden, but all existing establishments. A single firm man saved all; this was Lakanel, of the National Assembly. To him was entrusted by the assembly the great measure, the reorganization of the Garden, and its protection against the savage mob. It was now called the Museum of Natural History; twelve chairs were named, and one given to Etienne Geoffroy.*

When appointed to the chair of zoology, Geoffroy was profoundly ignorant of that study. He was a mineralogist. But he speedily overcame the difficulty with the assistance of Daubenton, who pointed out to him that real zoology was still to create; and that it did not exist in France (nor any where else), as a science.

"The Garden" in which we now find Geoffroy placed as a professor was first formed in 1626, and is due chiefly to Guy de la Brosse, physician to Louis the Thirteenth. About forty years afterwards it was first opened as

* His father's name was Gerard Geoffroy; this, at least, is his signature in a letter to Lakanel. When I had first the pleasure to meet him, I addressed him by the name of St. Hilaire; but he explained to me that his true name was Etienne Geoffroy.

a botanic garden for the growth and exhibition of medicinal plants. It was a sort of apothecaries' affair, half trading probably, half scientific, like the thing at Chelsea; and in fact it became at first a school of pharmacy. But under Fagon, an amateur and relation of this De la Brosse, it lost its trading apothecary character, and became scientific. Botany was taught *scientifically* * by Fagon himself, and in succession by Tournefort, Vaillant, Jussieu. Human anatomy was introduced and taught by Duverney, whose eloquence excited the envy even of Voltaire. He was succeeded by Dionis, Winslow, Ferrand, Vicq d'Azyr. In my own time, Portal was there the patriarch of anatomists. Chemistry was lectured on first by Rouelle and Macquart; Thenard and Vaucquelin taught chemistry in the gardens in my time. The Garden escaped at last out of the hands of the King's doctors, and was entrusted to De Fay, a young man, Member of the Academy. He was succeeded by Buffon.

Buffon was not a naturalist properly speak-

* As Linnè, Jussieu, and last and greatest, Robert Brown, had not yet appeared, for *scientifically* we must, I think, read *dogmatically*.

ing, nor was he a scientific man in any sense of the term. He was a literary man, a theorist, a profound thinker on a few facts, but he must not be compared with Goethe, the most extraordinary scientific man that perhaps ever lived, excepting Aristotle. He cared not for demonstrations; and although Daubenton worked with him as an anatomist, he took no interest in his labours. Objections were even raised to Daubenton's dissections, and they were omitted from some editions of Buffon's works. In fact, had Cuvier not appeared soon after, true science, in the hands of Mertrud, Geoffroy and others, must have gone to the wall; so that, but for the accidental birth and success of this great man, natural history and zoology and fossil bones would have fallen back into the hands of speculative naturalists, amateurs, and triflers, handicraftsmen, and, worse than all, literary men, followers of Smellie, of Goldsmith, and St. Pierre. The stern rigid "demonstrator" placed an eternal barrier to such a relapse—never, we trust, again to return. But I venture not to predict. Man's intellectual and social history is a circle, not a line, curved or straight, onwards towards a point. New modes of civilization arise, new social institutions—new fashions in literature—new modes of expression—now it is all religion, and now all licentiousness; but the unleavened mass of mankind remains throughout all ages the same. So does the eternal truth: nor time nor circumstances, whilst man's organization continues as it is, can affect or depreciate the Homeric ballad—the Venus—the Cena of Leonardo—the Parthenon as it stood—the "Ossemens Fossiles,"—the Shakspeare drama.

Buffon died in 1788, and the Garden fell once more into the hands of the *laissez faire* men—the men who, not feeling their own deficiencies, think everything around them good and excellent. In political life they form the so-called officials or red-tape men; in literary and scientific life, they get into office by cringing, menial services to some one in power. To all progress they offer the most determined resistance in a body. They are men thoroughly embued with a debased spirit, who have contrived by disreputable intrigues to foist themselves into office. For a time this class seems to have had possession of the Garden of Plants; they are, if possible, worse than the same class in England and in Holland.

The Celtic *laissez faire* man could not even extend a mere *collection* (for there were no museums there any more than in England), of which the nucleus had been bequeathed to them by Buffon and Daubenton. They had accepted the ideas of Buffon, and proposed to stereotype them as they had done those of Carl Linnè; before him, of Rondolet and Bonnet; before them, of Aristotle. At the fortunate moment came the great revolution, and swept for a time the reign of mediocrity from France.

The revolutionary assembly of France, even when menaced by all Europe or rather by the aristocracy of Europe, (for the peoples themselves were wholly with them, until Napoleon appeared in his true colours), deliberated, and organized institutions, and voted the means for their maintenance, which the wealthiest monarchies, in the midst of the most profound peace, had never even contemplated.

I know not to which of the great men whose lives I now give, ought to be ascribed the merit of having regenerated the collections of the museum. I feel inclined to ascribe the merit, wholly and solely, to Cuvier. But be this as it may, when they entered on office as professors, they found that the *juste milieu* men had been at work; science was at a stand-still, but the salaries had been paid as usual. The skeletons collected by Daubenton had been thrown into the cellar;* in 1793 there were in the zoological collection only four hundred and thirty-three individual specimens. The collection of Le Vaillant, in South Africa, had probably not been added to the museum.† But it was not a museum any more than the things in England of the same character; it was a mere collection. To imagine "a museum" is the gift of genius alone. The only *museum* I know of in England, is that formed by Mr. Hunter, now most unhappily and incongruously forming a part of, or incorporated with, the *collection* of the College of Surgeons! What has science to do with trading corporations? literature with guilds? art with academies? I will tell you; they present snug berths for the cast-off servants, and sons of servants of the men in power. That is all.

As England owes her most liberal laws and institutions to Richard III. and Cromwell, so France owes everything, as regards science, literature, and art, to the revolutionary as-

* See note II. † See note III.

sembly of 1793. The men of the "bourgeoisie" concealed themselves, or fled to Ghent; the aristocracy, priest, king, were destroyed; the money-lender held his tongue; the human mind all at once threw off its shackles, and men with giants' minds appeared. To enumerate them would occupy a volume. Amongst them was Lakanel, himself neither a hack literateur of the Guizot or Thiers school, nor a sham scientific man, like a Peel or Buckland, but a plain citizen. In one hour he carried through a measure unlike anything ever proposed in the first monarchy of the world. It was perfect, or nearly so. The results may be seen in the works of Monge, Bertholet, Laplace, Humboldt, Gay Lussac, Arago, Geoffroy, Cuvier.

It is asserted by Isidore Geoffroy, that his father, with Lamarck, created the Museum of Zoology. Be this as it may, I feel assured that in 1820-21 Geoffroy had ceased to labour therein, the whole direction being evidently in the hands of Cuvier. On the 6th of May, 1794, Geoffroy started as a lecturer; but I could not learn that he was ever popular or fluent. In 1820-21, he seemed to me to have ceased to teach; having become wholly unintelligible to the students, in consequence of his transcendentalism, but few attended him.

He is said also, by his son, to have created the Menagerie, and in 1793-94 saved the life of Daubenton. At this time Cuvier was at Fecamp, unknown. But some of his MSS. having been forwarded to Geoffroy, he saw in Cuvier another Linnè and predicted his future reputation. At that time, classification, individual species, and the manners and habits of animals (as if such things were of any moment to civilized men) formed the aim and object of Geoffroy's studies, as of all naturalists then and now. He urged his friends to bring Cuvier to Paris. Daubenton, older and more experienced, recommended Geoffroy, with the instinctive sagacity which marked his character, not to push on the fortunes and views of Cuvier so fast; Geoffroy would not or did not take the hint. Cuvier, destined to overshadow his friend, reaches Paris in 1795, he was then twenty-five years and a half old, Geoffroy twenty-three. For some time they worked together, and wrote joint memoirs, full of happy and great thoughts, shadows of the grand views which were to follow.

But France was now guided by the hand destined to act the grandest part ever played by man, in the history of the world; whose very name *has* become a talisman wherewith to conjure up armies of steel-clad men. Napoleon, at that time (1798) General Buonaparte, becoming an object of dread to the political cliques of Paris, the Thierses and Guizots of their day, had been appointed to conduct an army into Egypt. Amongst other great men whom Napoleon invited to accompany him, in the quality of a scientific staff, were Geoffroy and Cuvier. This was in 1798; Cuvier declined, Geoffroy accepted. This episode in the history of the life of Geoffroy merits from all thinking minds the deepest reflections. The time is come to write its history, and to contrast the doings of commercial, trading England, with scientific, literary, artistic France. But this is not my object, and to do more than mention it were foreign to the aim of this Biography.

For a similar reason I must touch but briefly on the episode in the history of Geoffroy's visit to Spain and Portugal. It partakes of the marvellous and the romantic. When in Paris another version was given me of the affair within the very circle of the Garden. According to his son, Geoffroy, by orders of Napoleon, plundered the scientific institutions and monasteries of Portugal, but so adroitly, with such urbanity, *politesse*, and kindly feeling, that the Portuguese themselves not only seemed insensible of the fact, but thanked him for the spoliation. Another version was given me in Paris, to which, however, I paid but little attention at the time. Geoffroy had for his companion an *aide-naturaliste*, described to me as a man of low birth, and unscrupulous. The plunder collected in Portugal by these *employés* of Napoleon "in the interests of science," amounted to seventeen large cases. Junot, to secure his own robberies, which were enormous, betrayed Lisbon into the hands of the English, and evacuated the place. The seventeen cases of plunder collected by Geoffroy by the commands of his master were for an instant detained by the English commandant. They were ultimately given up. On being opened, it was reported that three which were detained by the English commander, were found to contain anything but specimens of natural history, MSS., &c. Metallic specimens of shapes and figures, on which heretics like myself look with horror and pity, and other rich plunder, were said to have mainly composed the contents of these three cases. I give the story as I heard it. With the promise that my remarks on these two episodes in the life of Geoffroy must be very brief, I here present them to the reader, together with an outline of that great theory which haunted the mind of Geoffroy from twenty-three to the close of life.

When Napoleon was First Consul the executive of France, for the time being, resolved that a blow should be struck at England, "the great tyrant of the seas." Various circumstances pointed to Egypt as the field on which to place that advanced guard, that precursor army, aiming at India, then as now the grand source of England's wealth, the mine out of which she pays her armies, the never-failing billet for her aristocratic adventurers, who in search of gold to replenish their impoverished finances, and maintain their sinking estates and dignities and class, find in the systematic legalized plunder of India a never-failing resource. The *administrative* of France, aware of this, and influenced no doubt by other motives in which the *personnel* would of necessity fully share, selected Napoleon, then General Buonaparte, as the man to strike that first blow against the aristocracy of England, and of Europe.* A gallant and well appointed army, and noble fleet, left Toulon roadstead, under the man whose star led him to Austerlitz and Borodino, to set for a time at Mont St. Jean. As a purely politico-military expedition its results are known; although of mighty consequence to England, as to all the world, they need not be here further adverted to. It is to the part played by the scientific men of France, foremost amongst whom was Napoleon himself, and more especially to the thoughts and actions of that great man whose life and labours I now consider, that I would desire for a brief space to direct the attention of my readers. The influence of individual lives over the affairs of men has been, as I have already stated, singularly misunderstood; entire nations have not unfrequently a direct interest in misrepresenting the character of great men,—in making them appear what they were not. Historians, who chronicle events which the world calls

* Peninsular Wars, by Napier.

histories, and which some mistake for "philosophy teaching by examples," dwell on the passage of the Granicus, the Battle of the Pyramids, and of Austerlitz, as if Alexander and Napoleon were mere soldiers—Hannibals and Wallensteins; Marlboroughs and Bluchers (I mention only the dead); Soults and Clives — bloodthirsty soldiers, robbers and plunderers. Austerlitz and the passage of the Granicus were no doubt wonderful events. The one opened a road to Asia, and on the field of Austerlitz Europe fell before the genius of one man. But such doings, great as they are, form but a portion, and a small one too, of the mighty career of genius. Alexander was but a few days in Babylon when he visited the temple of Belus; conferred with its priests; sent copies of the chronological tables to Callisthenes in Greece; placed ten thousand labourers on the walls of Babylon, which had been injured by Cyrus. Let us take a brief view of Napoleon's first move in the cause of humanity, science, civilization, when fortune placed at his command the means of showing to the world, that war with him was the means, not the end—plunder the means, not the aim of that gigantic mind which first conceived, and all but executed the greatest of all human conceptions—the liberation of man from the curse of hereditary imbecility. But, before following this illustrious man to Egypt, let me first state that doctrine of which, if he was not the originator, he at least introduced to the notice of the scientific world; startling the minds of men with a deep glimpse of the past and the future.

DOCTRINE.

The doctrine taught by the transcendental anatomists of Germany cannot be formuled in so clear a manner as strict science demands. Nevertheless, it may be made perfectly intelligible to those competent to generalize their ideas, and to reason abstractedly on science. To the mind occupied with individual facts, disjointed details, or observations mechanically grouped together, the doctrine of unity, of the organization, and of all the mighty results it leads to, will for ever remain a mystery. Such persons, and they comprise probably the greater part of men, solve all problems in physics by a moral element, namely, a direct appeal to a supreme cause, of the nature of which they know nothing, and never can know.

In this short biographical notice it were out of place to discuss the principles of the theory at great length; this I often did with Geoffroy and Cuvier himself, so early as 1820—21. At that time the two friends had not openly differed, and by admitting certain of Geoffroy's views, and qualifying others, Cuvier contrived to avoid an open breach with Geoffroy. It came at last, as I then foresaw it would.

Whoever looks attentively at the structure of man and animals, especially if aided by anatomical research, may readily observe that, generally speaking, all vertebrate animals * are formed of organs strongly resembling each other, however remote from each other in the zoological scale the species or genera may be; that in point of fact it is obvious that all have been formed on one plan, or scheme. This plan, or scheme, must have existed at their creation; the scheme, or plan, must be regulated by secondary laws, such as those of attraction, for none else are intelligible to the human

* Animals having a cranium and vertebral column—or back bone; a most incorrect expression, calculated only to mislead.

mind. Aristotle, Leibnitz, Pascal, Newton, Harvey, all thinking men, of all ages, will admit this statement to be essentially true. But this view in a sense was yet far from scientific.

Towards the close of the last century scientific men, first in Germany, and then in France, proposed testing the doctrine by an appeal to observation; ascertaining the extent to which it might be carried; bringing it within the pale of strict philosophy, and with it zoology, which had hitherto been undeserving the name of a science.

It was argued, though not from the observations I here offer, many of which are peculiar to myself, but from others similar in their nature, that in some animals there existed structures seemingly perfectly developed, whilst in others, the same structures might also, with great care and nicety of observation, be discovered in a rudimentary, or undeveloped condition. The conclusion drawn from this was, that one plan existed in all, and that even although such vestiges could not be found in all cases, it was not unreasonable, nor even unphilosophic to conjecture that such existed, either microscopically, or under some other unrecognizable form.

Thus was first proposed the great doctrine of Unity of Organization, for at least all vertebrate animals. As illustrations I shall mention the third eyelid, perceptible enough in man, though clearly a vestige; more developed in the ox, horse, dog; still more in the elephant; most of all in the bird,—ever the same elements nearly, are found in all; it is merely a question of size and function, but not of kind or organization. Or take the cartilaginous skeleton of the nostrils; small, beautifully formed in man, and especially in woman; the muscles rudimentary, the protector cartilages of the great chambers of the nose, which divide them from the vestibule, scarcely apparent; the muscular apparatus, and many of the remaining cartilages all but vestiges—that is rudimentary. Now look at those in the horse; lastly in the whale, where the protector cartilages have attained their maximum of development;—of enormous size, they fill up the apertures leading to the nasal chambers, when plunging deep the whale seeks in his flight the unfathomable depths of the ocean.

The foot of the horse is always prone, and yet a rudimentary pronator muscle exists. These facts, and thousands of others, which support the grand doctrine of Unity of the Organization, cannot be overthrown, cannot be denied.

But a question arose; how far can they be carried? must vestiges always exist? are the primitive germs identical, and equally numerous? and if so, why do the vertebræ differ so much in number and form in different animals? What has become of the germs? have they disappeared, wasted away, shrunk to nothing, been absorbed by other structures, or turned to other account? All these views may be true, but the problem has not yet been solved satisfactorily. The plan of unity does not require that all the material germs of the organs be the same in all animals; it is sufficient that the plan seems one. The number of vertebræ may in one animal be twenty, in another two hundred; it is merely a repetition of a first principle; a repetition of the primitive type—the ideal vertebra of the scheme. In comparing then these vertebræ in different animals, we must be careful of our determinations; for whilst in a sense all the vertebræ are analogous to each other, their identity, called by the Germans their homologies, may be difficult to prove. What is third in one may not be the third in another animal, but the tenth. The numerical enumeration of the vertebræ is a mere apology for the incompetency of the mind to translate their meaning, to determine their nature and place, to read the structure aright.

It was now recollected that the embryo of all living things undergoes mysterious metamorphoses before its final development, which does not happen until long after birth. On re-examining these from a philosophic point of view, it was discovered that the embryonic forms resemble the normal, or persistent forms of animals lower in the scale; that the embryo of man, for example, has at first both gills and lungs, traces of which structure, the gills, I have seen on the necks of persons grown up to mature years; another confirmation that at first,—that is in the embryonic condition,—all animals show the same forms, have the same organs, display the same plan. These embryonic forms do more than this; they prove that the varieties in human structures depend on the persistence of these embryonic forms, and that most monstrosities owe their existence to the same cause; and finally, that the human embryo shadows forth in the history of its growth, from conception to birth, the history of the forms of all that lives; lastly, of all that ever lived from the first appearance of life upon the globe.

It was the opinion of Geoffroy, as we have seen it was that of Cuvier, that there never had been but one creation. This, also, is my own opinion. I believe all animals to be descended from primitive forms of life, forming an integral part of the globe itself; and that the successive varieties of animals and plants which the dissection of the strata of the earth clearly sets forth, is due to the occurrence of geological epochs, of the power of which we cannot form any true conception.

We know not then the causes of the specific and generic differences in animals, nor why such differences continue fixed for a period—the historic period for example; they depend, no doubt, on secondary laws, which some future Newton may discover. For the greatest of all discoveries remains to be made; the causes, namely, the why, the wherefore of the varieties of living animals and plants which since the period when chaos disappeared and order commenced, have constantly decorated the earth, the air, and the waters.

Was there ever a chaos? I doubt it; the

dreams of Ovid and Milton may be poetry, but they are not science.

As science proceeds the links in the chain, or chains, of living beings are gradually being filled up. Already De Blainville has overthrown the generic and specific characters of the ancient elephants, rhinoceros, mammoth, on which Cuvier prided himself so much. The transmutation theory is a stumbling-block in the way only of those who will not see the truth. Nature left no gaps in her grand scheme; the gaps referred to simply denote the narrow character of human knowledge. Unity of design implies in this instance unity of execution: if gaps appear, the time is not come for their being filled up. In the fulness of time all will be developed, and then, and not till then, if ever, can we comprehend the great scheme of creation.

Although the embryo fish of the present day resemble the adult forms of the past, we must not infer that it is more perfect; each is perfect after its kind and time. They live in distinct geological epochs. The primitive fossil forms imply gigantic structures, with a robe, or external covering, at least equal to what now prevails. If man appeared last, of which we have no proof, it does not follow that he is more perfect than any other animal. But to him has been given divine qualities, by the exercise of which he places between himself and all other created beings a gulf they cannot pass. Universality, the type of immortality, resides in him; in all men to a certain extent; in some it is transcendant. It formed the leading feature in the minds of Aristotle and Bacon, the sculptor of the Venus, the planner of the Pyramids, the architect of the Parthenon, and it pervaded the grand minds of the men whose lives I now consider.

It is right to caution the student of the transcendental in anatomy that much vagueness of expression and doctrine prevails throughout the works of Geoffroy; he told me that he never wrote but under the influence of inspiration, and I firmly believe it. When I first saw him he was intently occupied in determining the true character of the opercular bones—that is the osseous gill-covers of fishes. He viewed them as the identical *ossicula auditus* or small bones we meet with in the cavity of the tympanum in mammals; that is, to use a German phrase, they were the homologies of these bones. The determination is one of great difficulty. His German student, Spix, went further, as may be seen in his great work, the "Cephalogenesis."

Shortly afterwards Geoffroy published his beautiful memoir on the "Ideal Type of a Vertebra." The vertebra he made the type of the skeleton, including in its full development the skeleton of the limbs. This is not the place to trace these doctrines further; my own opinions have never altered since 1821, as may be proved by my writings and lectures. They are essentially based on a rigorous determination of the value of each structure or organ in every animal, and a detection of the strictly corresponding or identical organ, that is, its exact counterpart in others; and this I imagine was, after all, the real object of all Geoffroy's researches. Of his paleontological views I have already spoken; their simple exposition, or rather the exposition of Burmeister, translated anonymously, startled the reading world a few years ago. "The Vestiges of Creation" raised the veil for a moment from the gaze of the great world, permitting it for the first time to look back into the history of the globe. Let us now accompany Geoffroy to Egypt; for it was in this mysterious land that the great doctrine of the unity of the organization took full possession of his mind.

GEOFFROY IN EGYPT.

We have seen that already in 1795, and at all events in 1797, Geoffroy was sufficiently master of the grand idea of unity of organization, as distinct from the larger generalization, unity of plan in the creation of organic beings, to express with tolerable precision his conception of the theory. In 1798 an event occurred which unquestionably exercised a strong influence over the scientific future of Geoffroy, urging him strongly forward in the path he had chosen. Berthollet, a name illustrious in science, waits on Cuvier and Geoffroy at the Garden, the bearer of an invitation to accompany Buonaparte in a distant expedition. Fortunately perhaps for science, Cuvier declined the invitation; Geoffroy accepted it. But for this, the "Leçons d'Anatomie Comparée" might never have been completed; the "Ossemens Fossiles," that logical and unanswerable demonstration of the true history of the earth, might have seen the light as a feeble argumentative memoir, open to numerous exceptions. The time indeed had not arrived when the history of the primitive world could have been written; the history of the existing Fauna was still to write. On this work Cuvier was, at the moment I speak of, employed. The anatomy of the fossil kingdom (organic) was but the sequel of the anatomy of that now existing. They are parts first and second of one great chapter.

As Cuvier followed Daubenton, who with a feeble light had already begun to explore this mysterious field, so Geoffroy followed Goethe. The illustrious German, the man of universal genius, had already explored this field, on which Geoffroy was ultimately to glean so successfully; but his labours and vast conceptions taking the world prematurely and by surprise were wholly misunderstood, neglected, and forgotten. He had appeared before his time. A link in the chain of evidence to render his writings intelligible to the mass of men was still wanting; that link was supplied by Cuvier.

As with Goethe so with Geoffroy; whilst he wandered in the land of the Pharaohs, dreaming of the past, his immortal friend Cuvier, an obscure anatomist, in his retirement in the Garden, unnoticed, was extracting from the mortal remains of animals usually despised by men, that element of power, the truth, destined to revolutionize the scientific world, and to level so disastrously the great bulwark of superstition, as to render its reconstruction impossible.

"The Expedition to Egypt" was wrapt in profound mystery; the destination of the army collected at Toulon was long unknown even to those who held the first rank in the army, and in the scientific commission which was to accompany that army. "Come with me," said Berthollet to young Geoffroy, "you shall have for companions Monge and myself —and Buonaparte for general." Such was the invitation. But a portion of the secret gradually oozed out: the choice of books for the commission revealed to a certain extent the object of the expedition, and April 1798, found Geoffroy and his colleague, the illustrious Savigny, equally prepared to explore Egypt and Syria. Two months sufficed to collect in the port of Toulon thirty-six thousand soldiers, ten thousand sailors, and a host of men already or about to be distinguished in literature, science and art,—Monge, Fourrier, Malus, Berthollet, Dolomieu, Geoffroy, Larrey. How this army sped, how it conquered, and was at last overthrown, what fate the fleet which carried it experienced, are now parts of history. My own remarks I shall confine to Geoffroy and the scientific portion of the expedition. In a military and scientific point of view, the campaign is said to have been alike honourable to the rival nations who contended for the mastery of Egypt. Every question has two sides; Napoleon broke in pieces the frightful government of the Mamelukes, we restored it; the gain to humanity in this I am at a loss to conjecture. We are fond of restoring ancient monarchical dynasties, and of creating them where they did not previously exist. Holland for example, and Greece.

On the 19th of May, 1798, this noble army, commanded by another Alexander, quitted the shores of France. The restless mind of Geoffroy, his love of scientific inquiry seeks for objects of research even on board a frigate. By turns the student of the engineers on board, by turns their instructor in the elements of natural history, the capture of a shark, on the 20th day, enabled Geoffroy to exhibit to the delighted crew the effects of galvanism, at that time a novelty in science. In the mean time, the flag of Nelson was looming in the Mediterranean; his redoubted attack was dreaded, and every effort made to escape from his fatal pursuit. Their efforts succeeded. Called to another destiny, Geoffroy narrowly escaped drowning, having visited in an open boat the officers of another vessel. Reaching Malta, the impregnable city is handed over to them by the "Knights." On the 18th of June, a day fatal to France and to the Empire, the fleet quits Malta; on the 30th of the same month they discover Pompey's Pillar. The army was landed that evening: such was Napoleon.

In the terrestrial paradise of Rosetta the commission of the sciences found a pleasant retreat; here they first established themselves. But the city was a desert; provisions scarce; servants still more so. The commission acted as cooks in their turn; when Geoffroy's day came "the company had but sorry fare." But the difficulties which usually beset all first attempts gradually yield; order is established and with it abundance. Already, in a month, the commission is called to Cairo; Egypt in the interval had been conquered, the law and order restored, and all the blessings of peace. With matchless energy, instinct, and judgment, Napoleon foresaw everything. Already he had issued instructions for the formation of the Institute of Egypt; the labour which costs the minds of a nation ten centuries to imagine he performed in ten days. The Institute was composed of seven members, representing each the arts and sciences; Caffarelli and Andreossi were members, so were Desgenettes and Geoffroy; Napoleon was *elected*, at his own request, a member and vice-president by the votes of the seven. Thus did he ever know what was due to all men. With him scientific men were not "humbugs," "quacks," "impostors," the terms usually applied to them in England.

Already Cairo had become a centre of civilization and letters. There sat an Institute; at the quarters of the General-in-Chief was a matchless re-union of talent; there they could listen to the words of the first of men. Geoffroy had the happiness to be chosen as the companion of the future Emperor, in his excursions. It was in the gardens of Esbekiah, and again as he was about to quit Egypt, that, conversing with his staff, these memorable words escaped him; they were addressed to Monge,—"I find myself here, conqueror of Egypt, marching in the footsteps of Alexander; but I should have preferred following those of Newton." But Monge replied that Newton had exhausted the field of discovery in physics, leaving nothing to those who might follow. "By no means," was the remarkable reply of Napoleon; "Newton dealt with masses of matter, and with their movements; I should have sought in the atoms for the laws by which worlds have been constructed." Thus was his genius universal.

Exploring every part of Egypt, Upper and Lower, reaching the shores of the Red Sea, dissecting and observing, composing memoirs for the Institute of Cairo; these were the daily and incessant labours of Geoffroy in Egypt. It was in this country, inspired by so many reminiscences of past history, that he first meditated his great theory of the unity of the organization, his doctrine of analogies and homologies, his theory of creation. But as yet his views were mere hypotheses; they required a demonstrator; now that demonstrator quietly laboured in France whilst all around was turbulence, violence, rapine, murder. In Egypt the calm continued about a year. At Damietta, in the lake Merzalet, he discovered and described the fish called heterobranche. An expedition to the Cataracts produced as a result the great work on Egypt; an immortal work, which does honour to the name of France.

And now in January, 1800, news reached Cairo that El Arich, the key to Egypt on the Syrian side, was in possession of the Turks, and the evacuation of Egypt, had been agreed to and signed by Dessaix and Kleber.

Geoffroy repairs to Alexandria, expecting hourly an order to embark for France, when an insolent message from the English General induced Kleber to tear to pieces the convention. The battle of Heliopolis redeemed the massacre of El Arich, and Egypt was once more reconquered. The scientific commission was to return immediately to France, but the embargo established by Sydney Smith continued for twenty months, detaining the party in Alexandria.

Kleber was dead, and the wretched Menou commanded. The plague raged in the city; for twenty-nine days Geoffroy, attacked by the endemic ophthalmia, lost his sight. Then came the horrors of the siege of Alexandria, on which we need not dwell. He returns once more to Cairo, where the brothers* meet. He again resumes his scientific labours with great success. Returning once more to Alexandria the commission, now in the hands of Menou, a brutal soldier, without a spark of intellect or taste, is all but broken up: insulted, neglected, despised, Menou demands from the commission their collection, which they refuse. Embarked on board the brig "Oiseau" they are constrained in miserable suspense to pass many weeks on board the vessel. Menou seems to have lost his senses. He had given orders to a French frigate to fire on the unfortunate brig, which contained the scientific commission. The English admiral interfered.

Why should I extend this episode? the principal events have been told. The English commander endeavoured to seize the collections of the French savans: they were saved to France by the moral courage of Geoffroy, who threatened to burn them should such a demand be persisted in. Nevertheless some articles

* Geoffroy's brother was an engineer.

found their way to England, which were originally intended for France. The trilinguistic stone of Rosetta, for example, which lay for twenty years buried in the vaults of the British Museum unnoticed, disregarded—despised! Its history is still to write. I leave to some future historian of the races of men the task of unravelling the mystery, as singular at least as the Asian, how the most remarkable monument of antiquity, the possession of which nearly led an army, small though brave, to risk again in a pitched battle their existence and their honour. Now this much-prized monument, on being transported to London, was held to be of no value—misunderstood—forgotten—despised! —The explanation of these curious facts I leave to the future historian, who may undertake a second edition of the "Mœurs des Nations et Peuples;" he will find the explanation, if I err not, in the history of the race. A few years ago there lay rotting in the same cellars the only specimen of the head of the Mysticetus which existed in Europe. Cuvier discovered it in the cellars or vaults of the Museum, where no man would have thought of looking for it. The Elgin Marbles, as I have been told, occupied for a time the cellars I speak of.

Things like these are wholly beneath the notice of a great commercial people; the foremost, notwithstanding, in their own estimation, in literature, science, and art!

GEOFFROY IN THE IBERIAN PENINSULA.

Iberia is the land of romance and of adventure, and such Geoffroy and his companion found it to be. Touching the character and completion of Geoffroy's mission to Portugal calumny has not been idle—it existed even within the walls of "the Garden" itself. The reports I heard affected him and his *compagnon de voyage*, an *aide-naturaliste* of the establishment, as they are called; young men, often without education, or but slenderly educated; adepts at bird-stuffing; good collectors; unscrupulous; political. Accompanied by one of these *aide-naturalistes*, M. Geoffroy left Paris at the bidding of Napoleon. The directions he received were to plunder Portugal of whatever she might possess of value or interest in science. It is probable, nay certain, that Napoleon, who never mistook his man, selected the person best adapted to carry through an unpopular measure. From documents published by the son of Geoffroy we learn that he executed his commission with consummate skill.

The museums of Holland and Belgium, Austria, Prussia, had already been plundered: there remained but Spain and Portugal. Lisbon and Portugal were assigned to the cares of Geoffroy. Let us follow Geoffroy into Iberia, that country which proved so disastrous to Napoleon; which wasted and consumed his armies; distracted his attention, absorbed his means. A French army, under the orders of Junot, and of no great numerical strength, invades Portugal; and in a month the French occupy Lisbon. This was on the 30th of November, 1807. Speedily, and without the loss of an instant of time, Geoffroy is named by the Emperor to proceed to Lisbon in "the interests of science." The written terms of the first decision of Napoleon in respect of the scientific foray were simply that Geoffroy "should visit the collections of natural history in Lisbon and its vicinity, and determine what objects could be usefully transported to Paris." But so soon as Geoffroy had accepted the commission "it assumed the importance not previously thought of." "It was extended, in fact, to all that could interest not only the sciences generally, but letters also and arts." "By instructions given partly in writing, partly oral or confidential, powers (to plunder) almost unbounded were given to Geoffroy."

How he executed this commission is best known to those who employed him: it was the old game of Egypt repeated on a smaller scale. Lisbon proved a kind of Alexandria in the circumstances almost parallel. Junot proved but little better than Menou. Lisbon surrendered to *les Anglais*, whose redoubtable fleet conveyed the French troops to France. It might have been well for Napoleon had he ordered Junot to be shot on his return; treachery was fast ripening around the great hero—his own staff were fingering English gold, and had become deeply enamoured of it. A Celt is a brave man, but you had better not trust him too far. Bourmont and Grouchy completed the catastrophe at Waterloo. But I must return to Geoffroy and his delicate mission to Lisbon.

Soldiers and sailors are simply robbers; they have no scruples, and no one blames them for so doing. But how was Geoffroy to act under such circumstances? He was no robber, but a noble, generous, kind-hearted man. I gather from documents which have been published that, laying it down as a principle that "the sciences are never at war," he resolved that his mission should be useful to Portugal as well as to France. A difficult task he had no doubt to perform; to plunder, yet to convince the plundered that he was not robbed.

He left Bayonne the 20th of March. Murat had not yet entered Madrid. The French had not yet been openly assailed, but lowering clouds were collecting in the horizon which predicted a coming storm. On the frontiers of Spain he learns the tidings of the abdication of Charles the Fourth, the occupation of Madrid by the French, the re-establishment of order. But all this was hollow. Ferdinand had proceeded as far as Burgos and Vittoria to meet Napoleon, then to Bayonne, where he learned, for the first time, the intended treachery. In a few days all Spain was on fire; then was raised in Madrid the terrible cry, *mort aux Français*, which ceased only at Waterloo.

It was at this very moment, when French blood streamed in Madrid and Spain, that the two naturalists peaceably and tranquilly journeying on from Madrid to Meajadas, between Truxillo and Merida, that their carriage was arrested. What was to be done? The mob insisted on their being put to death: they were Frenchmen; that was enough. Geoffroy, acting instinctively, decided on making an attempt to reach the Portuguese frontier, which was but a few leagues distant, rather than attempt a return to Madrid, which was remote. At this crisis of their affairs an escape presents itself. The master of the inn at whose house they rested happened to be a Frenchman, long settled in Spain, and in close connection and friendly intercourse with a band of smugglers, and he recommends to Geoffroy these dangerous friends as guides during a night march, by which they hoped to reach Portugal. But new difficulties arose. A civil officer arrives at Merida and arrests them: they are thrown into prison at Merida.

For three days the travellers remain in this frightful state, hourly expecting death; a mob threatens the gates of the prison. When everything seemed desperate, their safety was at hand. It had happened some days before, whilst travelling in Spain, they overtook a carriage which had been overturned on the road. This carriage contained a Spanish lady, who was slightly hurt by the accident. With the gallantry of the Celtic race, Geoffroy and his companion placed their carriage at the command of the lady and her family, who accepted it; our travellers followed the carriage on foot to the nearest town, and left them only when they could no longer be of service to them. This adventure saved their lives.

The lady was a native of Merida, wife of an officer of rank, and niece of the Count of Totrefresno, governor of Estramadura. She learns in Merida the danger to which Geoffroy and his companion are exposed, and she determines, if possible, to save their lives. At midnight of the 11th and 12th, by the authority of the governor, the gates of the prison are thrown open, their carriage restored to them, an escort appointed, and on the 13th they reach Elves, in Portugal, then held by General Kellerman.

It is lamentable to add that this noble and generous action ended fatally for the Count de Totrefresno; suspected of treason, he died a victim to his goodness of heart.

Arrived in Lisbon they found Junot installed with supreme power. An able governor, indulgent to the Portuguese, rigorous but just to all. And now commenced Geoffroy's labours, on which I need not dwell. Junot, his ancient *compagnon de voyage* in Egypt, issued orders that all scientific, literary, and artistic establishments should be open to Geoffroy. The alarm which naturally spread abroad that the collections were to be plundered in "the interests of Paris," soon subsided on Geoffroy's assurance that he visited them simply as "a visitor," for his own especial study and information. "The museums, libraries, convents, and royal palaces will alone be visited (Junot's orders seem to have extended to private houses); exchanges will be made, gifts presented; nothing shall be removed by violence, all by conciliation."

These plausible assurances restored the confidence of the Portuguese; nevertheless, Geoffroy visited some private mansions (particularly emigrés) of the wealthy who had fled their country. He first visited the Convent of Notre Dame de Jesus, and acted with great moderation. He there claimed some fossil remains *of no use to them*, and some crystals in duplicate. The museums of natural history and the libraries were plundered extensively, but all in "the interests of science." His friends assure us that he brought with him from Portugal, in addition to seventeen enormous cases of objects of vertu and utility, the good will of all honest men.

I must not pass over a most amiable trait in the character of my esteemed friend. The botanist, Brotero, lived in poverty and obscurity at Coimbra; Geoffroy presses his suit with Junot, and obtains for him full compensation and relief. He who saved the life of Daubenton, of twelve priests, of two archbishops, had a noble and generous heart, ever ready to succour the afflicted. Verdier, also, whom the Duke of Abrantes disliked, was restored from exile at the pressing instance of Geoffroy.

But now the 21st August had arrived, and Vimiera decides the fate of Portugal—an armistice, the convention of Cintra, and the evacuation of Portugal, were the result.

At the moment of embarkation the old struggle of Alexandria was about to be repeated; he was not wrong—he received an order to abandon all his collections. Nothing daunted, he tries *negotiation* and succeeds; the English commissaries agree to share the plunder with him, and a third is allotted to Geoffroy as a personal favour; not accorded to France but to Geoffroy. Returning to the charge, he at last obtained from the generosity of the English commander the most favourable terms: he was called on to leave only four packages, and permitted to remove the remainder to France.

With Junot he embarked in the English frigate, the *Nymph*—they reached Rochelle early in October, and were shortly in Paris. I have heard some strange stories in Paris, in the Garden itself, of the contents of the boxes removed from Lisbon, which stories I do not choose to repeat. When the assembled nations of Europe, in 1815, demanded from France the restoration of the pillage of so many capitals, Portugal was silent. The name of Geoffroy is, we are assured, greatly esteemed in the Peninsula.

CONCLUSION.

In 1808, on his return from Portugal, the Chair of Zoology, in the Faculty of Science, just added to the University by Napoleon, was offered to him, and accepted. Thus was he the first Professor of Zoology of the Faculty, as he had been of the Garden, or Museum. Towards the close of 1809, he commenced a series of lectures strongly tinctured with transcendental views. His programme here had no limits but science itself. We have the testimony of Dumas, that from this chair he first made known to France the doctrines of philosophic anatomy. In 1814, the European races of men, roused to a sense of their danger, marched on France: 1815 followed: the result is known. Geoffroy took a part in the political struggles of the day, but did not compromise himself. He held his official appointments unmolested to the close of life, on the 19th June, 1844.

The most remarkable event of his life was his struggles with Cuvier, before the Institute of France; the first in 1830; the second in 1832. Goethe was present, and the scientific world looked on in suspense. Cuvier's last attack on the doctrines of the Transcendentalists was made from the chair of the College of France, on the 8th May, 1832; five days afterwards, Cuvier had ceased to live.

Geoffroy's opinions never changed. He belongs to a future age.

PART II.

SECTION I.

Leonardo da Vinci,

HIS LIFE AND LABOURS.

In the preceding sections, treating of the life and labours of Cuvier and Geoffroy, the author of this biographical and philosophical study, has examined the relation of anatomy to science and philosophy; for to discover the true relation which this instrument of thought exercises over man's knowledge of the busy world, past, present, and to come; or, in other words, to discover through its means, a history of life and a theory of creation, is the real object of the part just finished. In the Section he now commences, it is his object to investigate the relation of the same element to the Fine Arts, those at least of sculpture and design. Of all artists Homer was the greatest; but his was a living picture — a moving, acting panorama—a magic mirror which he alone possessed; and holding up to mankind they saw therein the actual world of thought and action. But the sculptor and painter, and even the dramatist, use other instruments—another mirror. In marble or on canvas the former represent the external world from one point of view; fixed, unchangeable—the materials they use admit of no more. In the external world they look for the beautiful, the perfect, the true. Homer knew all this, but so did the great men whose lives I am about to sketch. As the object of his labours was chiefly man, his form as expressing his mind, his thoughts, I naturally here inquire into the influence which a scientific knowledge of man's mysterious and wonderful interior may have exercised on art—art, whose object is to represent only the exterior—that decorated surface presented to man by Nature; or, in other words, to ascertain the true relation of anatomy to art—that science which, in the hands of Cuvier and Geoffroy, as we have seen, revolutioned the thinking world, revealed the history of the globe, and threw a light—dim I admit,—upon the secret of Man's creation.

The following inquiry into the true relation

which anatomical science bears to art, was undertaken chiefly with a view to terminate a controversy which has prevailed for at least some hundred years. The matter in dispute was, "the importance of a knowledge of anatomy to the artist:" the relation, in fact, of anatomy to art. The author of this inquiry had long been convinced of the unsoundness of the views of West, Bell, Haydon, and the Anatomical school of artists generally, wherever they may be. He did not question the utility of a knowledge of anatomy to the artist, but he questioned altogether the present mode of instruction, which in his view leads to a total misdirection of the artist's studies.

But it had been said, especially by Sir Charles Bell, "Michael Angelo and Da Vinci were the best anatomists of their day, and, therefore, they were the first of artists." In this assertion the author is forced to see two propositions perfectly distinct; these propositions he has examined separately in the present inquiry.

The sketch-book of the immortal Leonardo, the author knew to be in this country. From 1825 to 1850, he endeavoured in vain to examine for himself this remarkable work; the favour was at last obtained. Having ascertained that the work was in the Queen's private library, at Windsor, the author addressed a note to Mr. Glover, the Queen's Librarian, requesting permission to examine the Sketch Book of the greatest of all artists. Mr. Glover, with the greatest politeness, immediately obtained for the author the requisite permission. The result of a brief inspection of this remarkable work is given in the following Lecture. In the opinion of the author, the contents of Leonardo's Sketch Book close the controversy. But they show much more than was imagined. With Dr. Marx he thinks Leonardo one of the greatest of men; the first of all artists—a profound anatomist—inventor of true iconographical anatomy; and, perhaps, even of the descriptive. The Sketch Book, in transmitting to posterity as it does the thoughts of Da Vinci, expressed as an artist would express his thoughts, shows that Leonardo had never suffered his anatomical studies to mislead him for a moment as an artist; or, in other words, that the aim and end of anatomical knowledge, or the true relation of anatomy to art, as contrasted with the conventional and theoretical, were perfectly understood by him.

INTRODUCTION.

There has always existed a difference of opinion amongst artists and amateurs in respect of the utility of a knowledge of anatomy to the artist. Some maintaining it to be of no utility whatever · others affirming it to be the essential basis of all sound artistic knowledge. I mean, of course, in respect of the human figure, the most beautiful and most perfect of Nature's works; whilst others, holding a middle course, have thought that a knowledge at least of the superficial muscles,—a superficial knowledge, in fact, of what lies near the surface, was amply sufficient for the artist.

The object of the present section is to decide this question so important to art—to show the true relation of science, that is, anatomical science, to art; and, if possible, bring to a close a controversy which has now endured for at least some centuries.

I. Of the history of ancient Greece at the period when the finest marbles of antiquity were sculptured, little or nothing is known. When Greece was plundered by Rome these marbles were transferred from Greece to Italy and lost sight of for many centuries. They were disinterred in Italy shortly after the revival of letters in Europe. Some were discovered in Byzantium, and some in Greece itself. The only valuable originals I have seen are those in the Louvre and the mutilated fragments called the Elgin Marbles, now in the British Museum. I do not think highly of the other statues in that gallery. But of most others, wherever placed, we have casts and models of various excellency; from which some idea, though an extremely imperfect one, as I have been informed, and can well imagine, may be formed of the originals, of which by far the greater number are still in Italy.

If there be one fact better established than another it is this, that during the authentic *historic period* of Greek art, the Greeks were wholly ignorant of anatomy. How it stood in the ages preceding we do not know, and yet it were most important to know this, for these immortal works were not carved during the historic period but prior to it. Homer lived and wrote before the historic period: his writings remain. Let us, for an instant, consider what they teach us in respect of the principles of art; not of manipulative art, but of that divine perception of the beautiful in nature, and in man, without a knowledge of which fine art does not exist. His divine genius led him to the discovery of the canon of beauty long before it had been chiselled in marble—he placed it in proportion—in form — and in the emblems of youth: in these it resides. The age of Phidias, it is true, is tolerably well known, and the Elgin Marbles are *supposed* or *generally understood* to have been chiselled by this great master, and his school. Even here there exists a something conjectural,—a defective evidence—a something to fill up. But admitting that Phidias and his school sculptured these wonderful marbles, this does not bring us nearer the solution of the great question; who invented the Greek canon of beauty? Who discovered the absolutely beautiful and the perfect, and carved them in marble? This, after all, is the great question viewed historically; for the other, namely, the *means by which* that canon was attained, is, after all, a subordinate one, having reference to a difficulty through which genius could leap at once; for transcendant genius requires little or no instruction. It is the mind moderately gifted which benefits by that. The scientific men of forty centuries have failed to describe so accurately, so beautifully, so artistically as Homer did, the organic elements constituting the emblems of youth and beauty, and the waste and decay which these sustain by time and age. All these Homer understood better, and has described more truthfully than the scientific men of forty centuries. The first question may be decided by some future Gibbon or Niebuhr: the second I shall endeavour to solve.

Before I approach this question, permit me to make a few remarks on the pre-historic period of Greece; that era which seems to have produced nearly all the great men.

On looking attentively at the statues within my observation, I cannot find the slightest foundation for the assertion that their sculptors must have dissected the human frame and been well acquainted with human anatomy. They, like Homer, had discovered Nature's secret, and bestowed their whole attention on the exterior. The exterior they read profoundly, and studied deeply—the *living exterior* and the *dead*. Above all they avoided displaying the dead, and dissected interior through the exterior. They had discovered that the interior presents hideous shapes but not forms. Men during the philosophic era of Greece saw all this, each reading the antique to the best of his abilities. The man of genius rediscovered the canon of the ancient masters, and wrought on its principles. The greater number, as now, unequal to this step, merely imitated and copied those who preceded them.

During the philosophic period of Greece, when authentic history commences, the pursuit of anatomy was strictly forbidden there. This prohibition extended throughout the Roman empire (of which Greece was but a province), until, and even after, the revival of letters. But, virtually, this prohibition of anatomical studies had ceased some time before the birth of Leonardo and Della Torre, and from that time artists had it in their power to study anatomy if they thought fit. Some did and some did not, and the utility of such studies is debated even to the present hour. Anatomy is not a science, but merely a mechanical art, a means towards an end. It is pursued by the physician and surgeon for the detection of disease, and the performance of operations; by both to discover the functions of the organs; and by the philosopher with the hope of detecting the laws of organic life, the origin of living beings, and the transcendental laws regulating the living world in time and space.* Its study has been recommended also to the artist by those who had not discovered, or knew nothing of *forms*, nor of the canon of the Greek. The result has been the mistaking the dead for the living. On canvas we have death-like dissected figures; in marble, cold, frigid, lifeless *statues*. Look at the sculptures in the Great Exhibition, and ask yourself, how it is that so few of those marbles, single or in groups, rouse your sympathies and receive your admiration. I shall tell you. It is the almost total absence of that *life-like surface* which alone distinguishes the living from the dead; the Venus de Medici from ——. Art then at its origin, at its commencement, owed nothing to anatomy. Art, as it arose in Italy, and as it has existed since, endeavoured to assume a new position, to adopt another ally —science. One thing at least is certain, the canon of the Beautiful and the Perfect was already displayed to the Italian masters, in the remains of Antique Art; Niobe and her daughters; the Venus de Medici and of Melos;

* See Life of Geoffroy.

the Mars and the Apollo were not to be surpassed. They were disinterred at or about the time of Da Vinci. The question which it was for them first to solve was, how were these matchless remains to be read, or understood. The grosser minds of modern men, at least of the European mind of that period, a compost of the barbaric races of the eastern and western worlds, minds sunk in conventionalism, brutality, and the most deplorable superstition, could not at first discover "Nature in Antique Art." This was natural enough. But the great Italian masters made the discovery at once. They did not exactly copy or imitate the ancient masters; they studied their remains and tried to understand them; and it is to the nature of these studies—the means, in fact, of acquiring excellence in art, that this lecture is mainly directed.

II. All artists, I think, admit that there is but one school of art—Nature herself. By Nature, I presume, is here meant, the material manifestations of life, and the inorganic masses composing the surface of the globe. To read this book aright, as regards human forms, the artist has thought it necessary to study the mangled and dissected dead. He endeavoured to add science to observation, and mistook the aim of both. I shall try to show him how he has been misdirected in his studies. The intention was excellent, the result pernicious to fine art. In representing the external world, whether on marble or canvas, it must not be forgotten, that man, wholly savage or civilized, forms also a part of Nature's plan; and his works and labours on the surface of the earth, whether by way of fancied improvement, or real disfigurement, must also be taken into account. Science is an admirable thing in itself; and to know anatomy is a valuable acquisition to all scientific, to all educated men. But the aim and end of its study must not be misunderstood. A knowledge of the interior of man's structure is essential to the surgeon and physician, to the zoologist and to the transcendental anatomist; it furnishes to the artist, as its highest aim, a *theory of art*. Hitherto, though not in all instances, it has unhappily induced the artist to display what he knows instead of enabling him cunningly to conceal that knowledge, as Nature has done, from the gaze of the world. He begins where he should end, and by drawing anatomically he displays that knowledge which he should keep in reserve merely to prove the correctness of his power of observing living forms.

In the history of the fine arts we meet with certain names, which, by the common consent of mankind, stand pre-eminent. The names are Leonardo da Vinci, Michael Angelo Buonarotti, and Raffaelo d'Urbino. Whatever be the merits of others, and they are unquestionably great, and occasionally transcendent, it is, nevertheless, all but universally admitted, that no names can be placed precisely on the same parallel with the immortal trio I have just enumerated. Whilst in some artists, we are pleased to own a knowledge of expression; in others an admirable tact and feeling for nature; in others a perfect love of truth and a power to express it, thus approaching in a restricted sense, and in as far as their other qualities admitted of, perfection. To none but to the above-mentioned three, favoured above all men by Nature, can be assigned those universal powers of mind to perceive, and hands to execute those grand perceptions of the external world. For as we never think of placing any name on a level with Homer and Shakspeare, or with Bacon and Newton, or with Alexander and Napoleon, unless it be the immortal Roman Dictator, so the names of Leonardo, of Angelo, and Urbino stand apart and alone.

It shall be my endeavour, throughout this lecture, to establish, in as far as I can, this fact; not merely as regards the history or reputation of these great men, but with a view to the establishing certain great principles in art: its relation to the human mind, to other arts, and to science.

Before I proceed with the elucidation of these principles I may as well remark here, that I do not consider the three great masters just named as superior to the men who carved the Elgin Marbles; who fashioned the Apollo and the Venus; the Dian and the Bacchus; the Farnese and the Laocoon: superior to such artists no man can be. They were perfect. Little or nothing is known of their persons; they belonged, perhaps, to the age of Homer; but be this as it may, the men I have now to speak of would probably have equalled them (surpassed them they could not) had they lived in the same country, and in the same age. What the starting point was to the great Greek artists of antiquity; what actually preceded them; what were the surrounding circumstances assisting or resisting their efforts to represent the "perfect" and the "beautiful" we know not. It is otherwise with the second or the Italian era of art. It is an historical period, and one over which history has thrown considerable light. With this period, then, we commence.

To understand the epoch we must return to the character and views of the races who overthrew the Roman empire, and with it, for a time, all civilization. Europe, and the noblest parts of Asia and Africa, had long been in the hands of a ruffian brutal soldiery, despisers alike of all forms of civilization: men who see beauty only in squadrons of horse and battalions of infantry. The coarse-minded Northern or Scandinavian had made his way deeply into Europe; the Celt, relapsing into barbarism, remained as fiery and fierce as ever. The Gothic and Sclavonian and Magyar races lorded it over the fairest provinces of the fallen Roman and Greek empires: the pseudo-civilization of the Saracen had been tried and failed. Northern and Western Europe continued in barbarism—hopeless, as it would seem, for this reason—that the races occupying the soil seemed unequal to invention. The soldier and the priest, and soldier-priest, had succeeded in extinguishing civilization. Men were no longer permitted to think; thus for centuries men lived like beasts of the field. Europe was a den of savage animals, many of whom had lost all traits of humanity but the human shape. The period was called "the dark ages." The western world was then in the hands of the Scandinavian, Gothic, Mongolian, Tartar, and Saracen races. The semicivilized Celt, retrograding from the period of the extinction of the Roman power, had lost that position in civilization which, even in the time of Julian, France, his head-quarters, no doubt enjoyed.

But as civilization forms a part of man's development, so in process of time Literature, Science, and Art began again to show themselves. First came Literature, emerging from the horrid abyss of dark and hideous ignorance. It assumed, as was to be expected, a monkish dress. Next came Science; last appeared divine Art, the crowning-top of civilization. They naturally reappeared in Italy, spreading thence into the various kingdoms of Europe. Of the progress of Science I need not speak, nor of Literature. As they progressed it was quickly shown, by a reference to the remains of antiquity, that before the advent of the northern barbarians a Roman literature had flourished superior to all which has followed, and that, prior to that literature, a still higher form existed in Greece. So it had been even with science, mechanical science, and with the natural sciences as they are called, or sciences of mere observation; the names of Euclid and Archimedes, of Thales and Aristotle, reveal a condition of mind we do not well understand. Art at last reappeared in Italy. What form it would have assumed had it not been influenced by the disinterment of the glorious antique statues it were impossible to conjecture. But be this as it may, their disinterment coincided with or shortly preceded the era of Leonardo, Angelo, and Raphael. The effect which the sight of the antique marbles produced on these great minds was such as to enable them to soar at once to the highest style of art; to look for and to discover and pourtray the beautiful; to paint and to draw the objects of the material world as they exist in Nature, and not as seen in the coarse minds of the mass; of minds ever ready to substitute their own miserable conceptions for Nature; to prune and fashion trees into fantastic shapes, to make artificial grottoes and waterfalls; to mistake large green-houses for palaces; to disfigure and deform that Nature they do not comprehend.

From this period art has never wholly become extinct. Retrograded occasionally she has, no doubt, but in the face of the remains of antiquity, Greek and Italian, it was impossible for true art to continue long in abeyance. The isolation of England for so long a period from the Continent during the era of Napoleon had an inconceivable influence upon the fine arts in that country from which she has not yet recovered.

The man whose life I am about to sketch is generally supposed to have been born in 1443 at Vinci, near Florence. The Italian Republics were then not quite extinct. He died in May, 1519, in France, at the Court of Francis the First, the rival and contemporary of Charles the Fifth. According to this account, he must have been about seventy-four at the time of his death, with all his great and divine faculties unimpaired. But others affirm that he died in 1518, and was born in 1452, which would make him at the time of his death only sixty-six. It were well for his countrymen distinctly to determine these dates —they are not without interest, as we shall presently see.

His master was Veracchio, whom he rapidly excelled. One fact seems certain—he preceded in time all the other great artists, so that Angelo and Raphael were acquainted with his works, for Raphael was born in 1483, when Da Vinci must have attained, by one account, his thirtieth and by another his fortieth year, and thus he must have executed most of his great works before the birth of Raphael. This divine artist had thus before him open to his contemplation the works of a perfect master; of one who had read Nature deeper than any one before or since. The immortal antique statues he also studied, and lastly the grand frescos of Angelo: yet Raphael *copied* no one—imitated no one. Creating his own style of art, that is viewing Nature after his own way, he also has left immortal works: second to none —such was Raphael. In claiming for Leonardo a precedence as to time, I do not wish you to imagine that such men as Angelo and Raphael imitated him; stole his ideas, or copied his works. Each followed that path which his divine genius dictated and inspired. Still the importance, even to the brightest genius, of contemplating the finished works of another equal at least to himself is undeniable, and its influence at times on his future labours inconceivably great.

Before Raphael was admitted to the sight of the then unfinished frescoes of Angelo, his style was comparatively tame and calm; but immediately thereafter his pencil and brush took a higher flight, placing him at once on the highest pinnacle of reputation.

The conclusion I would draw from the fact of both those artists, Angelo and Raphael, having seen Da Vinci's finished works when they themselves were comparatively young, is this: the perfect in form and the absolutely beautiful in Nature had been already revealed to them by the disinterment of the antique statues: in the drawings of Da Vinci, their master minds readily discovered the steps taken by Da Vinci to place on canvas, or on flat surfaces, outlines of form of matchless accuracy;—second, to represent Nature as she is;—third, to group or compose in such a way as to leave no doubt on the mind of the observer as to the intention or meaning of the artist;—fourth, to discriminate between absolute beauty as dependent on form alone, and other qualities which, from the poverty of language we also call beauty and beautiful; such as the beauty of youth, the beauty of expression. Some landscapes we also call beautiful, meaning pleasant to behold. The source of our delight in looking at these has not yet been clearly explained. When men depart from a strict analysis of their thoughts and sensations, and the reasoning mind usurps the place and functions of the instinctive, no absurdity will be found too hard for them to state, and to believe. Men like Voltaire, Alison, Jeffrey, maintained that there is no such thing as absolute beauty. Devoid by their nature of a *love of form*, and a power to perceive and admire its presence, to detect and regret its absence, they thought all beauty conventional. Utilitarian ideas of fitness and usefulness beset their material minds. Carry out your principles, you men of adaptation and fitness and utility; if what you say be true of form—if all forms be indifferently beautiful, conventionally beautiful, so must all colours; so must all music. But should it happen, as it probably was the case, that a vile Saxon ballad, a Dibdin song, seemed to you equal, perhaps superior, to the glorious conceptions of Beethoven and Mozart, you do not mean to dictate to those who have musical ears, or to maintain that your judgment in such cases is as good as theirs.*

The mind of Leonardo was altogether different from that of most men. He was an engineer, a mathematician, a machinist, a chemist, an anatomist, a physiologist, an artist, a philosopher, and he excelled in all.† The "Equestrian Combat at Pisa" was one of his early works. It is said to be perfect.

* The total absence of taste and fine feeling for all that is noble and great in man, may lead even the most laborious *littérateur* into errors highly ludicrous. Mr. Alison, in a preface to the "History of Europe," proposes estimating the social condition of France and England, respectively and comparatively, by the quantity of beef and mutton consumed in the two countries! By this estimate how meanly would ancient Athens and Corinth come off when compared with Liverpool and Wapping. Mr. Alison is, I believe, an Englishman.

† A lecturer on "Pictorial Anatomy" designates Leonardo as a Jack of all trades and master of none! I have heard the same remark applied to himself. The blame of such appointments, the electing to Professorships of Anatomy in Academies of Art, Surgeons, and Surgeon-Apothecaries, to whom the principles of art are unknown, rests of course with the patrons.

But so also was the "Cena, or Last Supper," of which some beautiful engravings have been made. Every time I examine this wonderful work I feel more and more struck with its absolute perfection. Notwithstanding his anatomical studies, of which I shall presently speak, and his love of truth, his love of beauty of form never escaped him. You will not find a coarsely made vulgar hand in any of the thirteen figures of that admirable painting.

As I one day stood intently gazing at this engraving, dictating some observations to an English lady kindly acting as my secretary, she called my attention to the shape of the bottles and glasses on the supper-table of the Lord. These, she said, could not have existed at the period represented. Her detected anachronism startled me from my reverie, and brought me back to this common-place matter-of-fact world in which we live. As I afterwards heard a similar remark made by a well meaning gentleman, who volunteered a few lectures on the principles of art in the Gower Street College of London, I shall here offer a few reflections applicable to all such remarks. When we examine with a view to criticize the works of little minds, of men of mediocrity, we look at once at *detail,* for there is nothing else to look at. An anachronism in time or place becomes a glaring defect. But when the works of the great masters are the objects of criticism, the best thing that a man of detail can do is to be silent. You cannot measure the great ideas of Rubens, of Raphael, of Leonardo, by such a rule. Your mind must either come up to the conceptions of the artist, or you must abandon the attempt altogether. These men, and many others whom I could name, show you the *unseen* through the *visible.* They show you that which alone can touch the deepest feelings of your nature, which is, in fact, the great aim of art. Compare Rubens' "Daniel in the Lions' Den" with Mr. Landseer's "Menagerie" and you will perfectly comprehend what I mean; or Teniers' "Quack Doctor," with anything else in the same style. You will then discover the difference between that which is perfect and beyond all praise, and that which is mediocre, or imperfect and open to criticism. But to do this on all occasions the mind of the observer must love perfection, and form, and colour; and these qualities are conferred on few. Of these few, Leonardo was one—perhaps the first, the greatest.

Whoever will compare the painting of "Macbeth and the Weird Sisters" of Zuccarelli, with the "Equestrian Combat at Pisa," the "Skittle Players" of Teniers, the "Village Festival" of Wouvermans, or the immortal "Sabine Rape" by Rubens, will or ought to discover that the same class of criticism is not applicable to all. What strikes you in Zuccarelli's "Macbeth" is, that the scenery in no way resembles the bare and blasted heath on which Macbeth met the dreaded sisters. You remark this at once, because the work does not rise to supreme excellence. If it did, the anachronism would escape your notice, or if pointed out by another would pass unheeded. It is with acting as with painting. Talma was perfect, and in any costume he would have been still the great Talma.* So

* I read a few years ago a critique in the "Times" newspaper on Talma in *Hamlet*; describing him as appearing in top-boots, &c. I need not say that there was not one word of truth in the critique. Voltaire's critique on Shakspeare was better; but, for once, the greatest of all critics encountered a work his conventionalities disabled him from comprehending. The difference being in this; Shakspeare saw man and the external world as Nature made

was John Kemble, even though as Prince of Denmark he was supposed to be walking in silk stockings and shoes on the bleak and stony shores of Elsinore.

With Kean or Talma on the stage all minor details ceased to interest. But when the acting is such as I saw it in the Princess's Theatre in 1848, the whole becomes ludicrous, and the weird sisters look merely frightful guys to terrify children withal. From the sublime to the ridiculous there is but a step; so said the man who excelled all others.

I have already said that Da Vinci had early discovered the true reading of the antique, that he had discovered "Nature in ancient art." With him there was no ideal, as modern artists and amateurs understand the term. He had discovered the true signification of external forms, discriminating them from internal shapes in as far as the artist requires to do. He understood practically the grand theory of the transcendental, the full develop-

them; Voltaire examined them as a courtier, a man of the world, full of common sense; of a sense of propriety; of a sense of the ludicrous. Even in the *épopée*, Voltaire's masterpiece, Shakspeare beat him; the "Troilus and Cressida" is vastly superior to "La Pucelle d'Orléans."

ment of which was reserved ultimately, as we have seen, for Oken and Goethe and Geof froy (St. Hilaire). But he did not remain content with this view. It seemed, by what I have next to relate, that he desired to know how far science supported his mode of viewing Nature. Here was the most dangerous step for the artist. Had there existed the smallest defect in his taste, his judgment, his feelings, from that moment he would have become the mannerist, the pedant, the mountebank; the anatomical dreamer. But it was not so. His great mind overcame even this. There is nothing in the history of art more curious than this trait in the life of Da Vinci I am now about to relate.

From whatever cause, Leonardo, in the prime of life and height of his reputation, associated himself with Della Torre, an anatomical teacher, a man of great merit and, perhaps, of genius. He is reported to have assisted Della Torre in his dissections, and to have designed the various organs of the dissected frame of man and animals through a period of ten years. When this fact was first mentioned to me by Sir Charles Bell, in 1821 or 1822, he coupled with it an observation which I, as early as 1810 knew, namely, that there existed in the library of George the Third a unique quarto volume of drawings, sketch MS. observations: the "Sketch Book," in fact, of the immortal Leonardo, containing his private thoughts, ideas, conceptions, views. This I gathered from the preface to Chamberlayne's work, professing to be a selection from the unpublished volume of Da Vinci's "Sketch Book." But as none of the engravings in Chamberlayne's work—those copied, I mean, from Da Vinci's "Sketch Book"—are, properly speaking, anatomical drawings, I still remained in doubt as to the exact amount of Da Vinci's anatomical knowledge. After long delay I at last, a few months ago, was permitted by the kindness of Mr. Glover, librarian to the Queen—sanctioned, no doubt, by the highest authority—to examine personally and for several hours a work probably without an equal in the history of design.

It is a small folio, prepared as a sketch-book, its leaves filled with figures, drawn by Leonardo, chiefly from dissections made either by himself, or conjointly with Della Torre. It comprises also some drawings of the vegetable world, and a few of machinery. But the figures are chiefly drawn from anatomical dissections, and in no instance could I perceive that Da Vinci ever mistook the dead for the living. As if to secure himself against the possibility of such an occurrence, he has drawn generally, and with a grace and spirit not to be surpassed, the living limb, with all its glorious exterior, side by side with the dead and dissected corse. He draws the dead as dead—the living as living.

In the same work are the drawings of the broad-headed horses ascribed to Julio Romano; a form of head which must, I think, have not only prevailed in Italy at the time, but been common near Florence. Turning the leaves hastily over I stumbled on a drawing of the semilunar valves of the aorta, in a variety of positions, so as to show their descriptive anatomy, and their physiological action. The corpuscules of Arantius have not been forgotten. Now all this occurred long before the age of Fabricius and Harvey; and even before that of Vesalius; for Della Torre and Da Vinci preceded all these.

It may have been that he was acquainted with the circulation of the blood. Who can tell the extent of his knowledge until the volume be carefully examined, figure by figure, line by line, page by page, by an anatomist?

It seems then to me, that Della Torre and Da Vinci were the founders of true Iconographic Anatomy, and perhaps even of the descriptive, re-discovered by Bichat. The text is written in Italian, in a cramp hand, and backwards; that is from right to left. It has never been published any more than the designs. Yet this man formed an era in science and in art. He shows their true relation to each other; the end and the aim of each; nor is it any longer a question with me, that had this work been published instead of his Treatise on Painting, the name of Da Vinci must have stood foremost amongst men; and art, instead of oscillating between conflicting theories and views, taken at once a fixed basis, safe from the pernicious patronage of common minds, whether aristocratic or plebeian.

There are a few peculiar circumstances recorded of the life and character of Da Vinci, which, though they may seem trivial, are yet full of meaning when rightly interpreted, and to these I shall now advert.

I.

His love of perfection was so great that his own works never satisfied him. Hence he was said never to have finished anything he undertook. He was more or less occupied with the portrait of Mona Lissa for four years; and this is said to have been his most finished production. But when we thus describe him as finishing nothing, it must not be supposed that he left anything imperfect; to him it so appeared; but to other men his works appear perfect.

II.

His acquaintance with Della Torre, an anatomical teacher, and, at the same time, a man probably of exquisite taste, biassed, no doubt, Da Vinci a little in favour of anatomical studies. But his paintings prove that he never mistook the end and aim of anatomy. What he precisely thought in respect of anatomy cannot be well known until his works be published. In addition to the folio in the Queen's Library at Windsor, which abounds with anatomical drawings, and descriptions in writing, no doubt of these sketches on the opposite page, Leonardo left fourteen or fifteen small volumes of MSS. and drawings. Of these, one I believe is still in Italy, in the Ambrosian Library; this volume must, I think, also contain some anatomical drawings; the remaining volumes are in Paris; they refer to a vast variety of subjects which Da Vinci studied, and of which he became master, such as optics, perspective, mathematics generally, machinery, engineering, etc. The only work of Da Vinci which has been published is his Treatise on Painting. It appeared in Italian and in French, long after his death. The French copy which I have in my possession, was illustrated by engravings from designs made by the celebrated Poussin. I can only account for this by supposing that before this work was published, the "Sketch Book" I have lately examined, had been transferred to England, otherwise it would have been quite unnecessary to call on a strange hand to illustrate any of Da Vinci's ideas.

III.

Had Da Vinci's works and sketches been published during his lifetime, or soon after, they would have formed an era, not merely in art, but also in science; and the controversy so long, and still carried on, respecting the position which anatomy holds to the fine arts, could never have existed. I can perceive in the oscillations of artists even now, between different opinions and styles, that the controversy still goes on, but under another form. Having of late become sensible of the beauty of form displayed in the antique marbles, and perceiving at last that there "is nature in antique art;" modern artists are too apt to forget the life-like surface which these ancient marbles also display. With these life-like appearances, which must never be omitted in sculpture, men's minds are generally familiar, and it is this which enables even those least familiar with art, to criticize modern statues without being aware of the faculty enabling them to do so. The truth is, that the mind of the observer sees in most modern marbles merely a *statue*, without the semblance of life. Compare one of them with the Ilyssus or Theseus of the Elgin Marbles, and my meaning will become apparent. On the other hand, certain artists, following Angelo, endeavour to give a life-like appearance to their figures by putting in action all the superficial muscles. The result is, an *anatomical study*—a *galvan-*

ised corpse. Follow Da Vinci. Draw the dead as dead—the living as living; never depart from truth. The dissected muscle, besides being dead, is quite unlike the living in form, and in every other quality.

IV.

The famous cartoon at Florence, representing an equestrian combat, was the work of Da Vinci. It is described as being perfect. When Angelo was requested to make another to match that of Leonardo, with the tact and forethought of a great mind he avoided repeating the subject of an equestrian battle. He saw, no doubt, that Leonardo's cartoon could not be excelled— perhaps not even equalled, and thus he chose for the subject of his cartoon an army surprised by another, whilst the greater number of the men were bathing in the Arno. The surprise and confusion arising out of such a circumstance, gave full scope to the grand pencil of the immortal Angelo, enabling him to hold his ground with his rival and predecessor. Thus he avoided the chance of a direct competition, and the still more dangerous charge of imitation.

To the man I have just described belonged exalted genius, an intense love of Nature and of truth, a power to perceive all objects in a new light; a love of the perfect beyond all other men. The lives and the works of such men form eras—that is, fresh starting-points for the human mind. Since Da Vinci and his contemporaries, there have arisen no such men; their age did not form them according to the Guizot theory; they formed their era. So with John Hunter. Surrounded by calculating tradesmen, vulgar-minded men, he struggles with them for his daily bread. Brought into daily, nay hourly contact with men whom he must have despised, he yet kept aloof, regarding them with scorn and contempt. He not only was not formed by his age, but in direct antagonism with it; his age, his contemporaries, his adopted country.

He overcame all, and left in his museum a monument like the Cena of Leonardo, to tell posterity, five hundred years hence, that great men are not formed by the times they live in, but the times by them. Men of mediocrity express the character of their times; they are its highest expression. Weak-minded people fancy such to be great men, merely because

Mirabeau, and Gibbon, and Canning are dead. This is the world; experience of the past teaches men nothing. Nor is this, perhaps, to be regretted; for were men continually to inquire into a knowledge of the past, attempts to discover the future would of necessity be given up; science would stand still; art be reduced to a mechanical imitation of antiquity; and literature be abandoned.

V.

Of the value set on the works of Raphael and of Leonardo by certain classes, we may judge by the following well attested facts. The monks to whom the edifice belonged, on the wall of which was painted the immortal Cena, cut a door-way through it, to suit their convenience! a nearer road by several steps, I believe, to their kitchen! So much for the class, priests. Queen Charlotte proposed cutting off the legs and feet of Raphael's cartoons to make them fit certain apartments at Kew! So much for the class, courtiers. Of the class, soldiers, I need say nothing. With them science, literature, art, are words without meaning. A long residence, it would seem in courts, monkeries, and barracks makes sad inroads upon the higher qualities of the mind.

VI.

There is an anecdote told of Raphael, the other great rival of Da Vinci, which tells us how perfect all Leonardo's works were esteemed, and proved to be. It was Raphael's wish to paint the Cena, but to avoid the composition of Leonardo. He attempted it and failed. He painted a square table, placed diagonally, but was forced to cancel his drawing, his mind telling him at once that no one could follow Da Vinci and attempt to improve on him. Learn from Raphael and Angelo by the value they set on Leonardo's works, those master minds whom none surpassed, that what is perfect cannot be improved. Study the works of the highest, but do not attempt to imitate or improve on them. The first will make you a mannerist, the second is sure to end in a ludicrous and disastrous failure.*

* The unpublished works of Da Vinci must contain a mine of artistic wealth, besides scientific. In the "Sketches" published, I think by Valvardi, and copied no doubt from the volume of "Sketches" by Leonardo, to be found in the unpublished volume still in the Ambrosian Library, there is a "Sketch of an Equestrian Female Figure," strongly resembling the "Amazon of Kiss."

In the life of Leonardo we have the first attempt made to discover and determine the true relation of Descriptive Anatomy to Art. With him rests the honour of the discovery, and of the just application of a knowledge of structure to the art whose object is to represent to man organic beings as Nature intended they should be represented.

SECTION II.

THE LIFE AND WORKS OF

Michael Angelo;

BEING A CONTINUATION OF THE INQUIRY INTO THE RELATION OF ANATOMY TO ART.

MICHAEL ANGELO BUONAROTTI was born on the 6th March, 1474, in the Castle of Caprese, in Tuscany. The original name was Canossa, for which the family adopted that of Buonarotti, or well-arrived. In accordance with the spirit of the age, his future reputation was foretold by astrologers. His genius was universal, — sculpture, painting, engineering, anatomy, poetry. But he did not succeed equally well in all. He considered himself merely as a sculptor, but he was never excelled by any one in painting, especially in Fresco. He was an architect of the highest order, and studied under Masaccio, who died at twenty-seven, and is much commended by Angelo.

It would seem then, by a comparison of dates, that at the time of Angelo's birth, in 1474, Leonardo da Vinci had already reached the thirty-first year of his age. By the time that Angelo had reached the twentieth year of his age, Leonardo was fifty-one, and had painted all his greatest works. Even if we adopt the later period assigned by some, Leonardo still must have reached the age of forty-two, Angelo being but twenty.

Angelo was preceded by a man who carried the art of painting to the highest possible perfection. Raphael was preceded by both. This important chronological fact must not be lost sight of, in estimating the respective abilities of these great men. Easy task. Each walked majestically in his own great and original path in life, guided by an inward light, and regardless of all around. Thus they each attained, though by different paths, a reputation destined to be immortal.

Angelo began life as a sculptor, at Florence. Ludovico, his great patron dies, and Piero succeeds. Piero viewed a sculptor merely as a mechanic or servant, but Angelo lived in the palace with him, as a relation. Young Angelo quitted Florence soon afterwards, in consequence of a ghost story related to him by Cordière. This story was as follows;—In the house of Piero de Medici there lived an Improvisatore, called Cordière. Lorenzo de Medici, the father of Piero, appeared to Cordière in a vision, and charged him to tell Piero his son, that he would shortly be driven from his home, never to return. This he privately communicated to Angelo, who recommended him to obey the vision, which Cordière hesitated to do. Lorenzo appeared a second time to him in a vision; with his friend Angelo, he proceeded to inform Piero of what had happened. Piero laughed exceedingly, with all his courtiers, at the story, passing many jokes at the expense of Cordière. Angelo and Cordière fled. Piero was afterwards driven from Florence, with his family, and never returned.

He next proceeds to Bologna. He was seized at his entrance into Bologna, and called on to produce his passports. You see how ancient this nuisance is, invented by feudal tyrants! His comrade having no passport, Angelo paid the fines. Throughout life his liberality and grandeur of soul were conspicuous.

His sympathies, like those of such men as Burns, were universal. To his new and wealthy patron he read Dante, Petrarch, and Boccaccio; these must have contributed greatly to the early formation of his character. Leaving Bologna in about a year, he returns to Florence, that city being again tranquil; and settles once more in his father's house! This father of Angelo was a nonentity, such as Cervantes has described. I ought to have mentioned that Michael began to draw when quite a boy; and he drew a bolder contour round the outline of his master; this sketch remains, and is the object of universal admiration. Like all great men, his docility or aptitude for acquiring knowledge, his powers of attention, and his educability were admirable. He sculptured at sixteen. One of his earliest works was a Cupid, which having been interred and concealed for some time, was then dug up and sold to Cardinal St. Georgia for an antique. This trick was told the Cardinal afterwards, who sent a message to Florence to arrest the seller of the figure. Angelo came to Rome with the messenger, but the Cardinal never forgave the trick. He recovered his money. He now worked in

Rome as a sculptor, executing many works. He cast figures in bronze; and thus, whilst still a very young man, he was already at the head of his profession,—a great master.

At every Court there are men of rank and influence, devoid of all taste. They are the natural enemies of men of genius — genius which they hate and abhor. They well know that men of genius can never become courtiers, nor bend to titled mountebanks, carrying gold or silver sticks, walking backwards like apes and jugglers before one of their frail fellow-creatures; hence men of genius are hated by all such persons; oppressed, crushed down, and, if possible, destroyed, or treated with sovereign contempt, neglect, or silence, which amounts precisely to the same in the grand struggle of life. Angelo met with many of these, but he generally overcame them. To allude in an especial manner to any such persons serves merely to bestow an immortality, unenviable it is true, on names which, but for such occurrences, must have remained for ever unknown. In matters of this kind we need not go so far back as the days of Angelo; Shakspeare was absolutely unnoticed by the court and world of fashion of his day.

Thus do courtiers, described by Montesquieu as men universally mean, servile, profligate, and selfish, sometimes succeed in despoiling a nation of its greatest minds. Whilst I now write, two figures have been set up at Southsea, under patronage of this kind, of which it is not saying too much, that they would disgrace any civilized country in the world. What must foreigners think of us when they see such things? But to return.

Angelo next painted in oils. His critique on Titian was most admirable. As a sculptor, Angelo was generally supposed to despise painting and painters, and men spoke of his private opinions, as if they knew them. To these conjectures in respect of his private opinions, he replied that he despised no form of Art; and conversing with the Pope, who had shown him some of Titian's early works, and requested his opinion thereon, he made the following very beautiful observation,—"If this young man would but learn to draw the human figure, he might in time become an admirable artist, for his taste is fine, and his knowledge of colouring exquisite." I find it stated in an excellent life of Angelo by Quatremère de Quincy, that it was at this time that Angelo's celebrated rival appeared. Now there must be some great mistake in this. Leonardo, as we have seen, preceded Angelo by many years. Perhaps the biographer means, that Angelo was now able to cope with Leonardo, and undertook, with his great rival, to decorate a hall at Florence. Each chose his subject, Leonardo preceding. The story and its results have been already told in my "Life of Leonardo."

When I contrast these noble doings with what I have seen in this country, I am naturally astounded.

The Cartoon of Pisa is now lost, but it was so perfect, that its mere contemplation made many great masters. Angelo saw Leonardo's works, and Raphael saw Angelo's, and studied both. This Cartoon was painted about 1503, when Angelo was only thirty; Leonardo must have been at this time at least fifty-two. They were not, then, strictly contemporaneous. Rivals they might be, and were. How great must have been the confidence and power of Angelo, to venture on painting a cartoon in the same hall which contained the finished production of his immortal predecessor! The drawing was executed in charcoal, and black and white chalk, and some of the figures were drawn with a pen. But he never painted or finished this work as a painter; it remained, therefore, a mere sketch or cartoon—but still superior, perhaps, to all the coloured paintings in the world. The infamy of having destroyed it, through malice and envy, has been assigned to Bandinelli.

So soon as Julius the Second ascended the Papal throne, he invited Angelo to Rome. But Julius and his artist could not determine on his first engagement, and so the Pope gave him an unlimited order to build a Mausoleum. This led to the origin of St. Peter's, in the planning of which Angelo was the architect. Some men, ingenious in inventing and maintaining paradoxes, have found in the building of St. Peter's the cause of the Reformation.

But in whatever light this may be viewed, certain it is, that the building of St. Peter's deprived the world of many noble works, which, but for this, would have been bequeathed to it by Michael Angelo. For in order to procure the requisite marble for this noble structure, he had to repair to the marble quarries of Carrara, where much time was lost.

Repeatedly, in the life of this wonderful

man of vast genius and exalted thoughts, there occur passages, showing that no mind so nearly touched the universal as his. He seemed to have thought nothing impossible. He had an idea of proceeding to the Sultan for the purpose of proposing to him some magnificent engineering plans. After having fled from Rome, he consented to return only when Soderini agreed to send him back as Ambassador from Florence.

About this time the French had possession of Milan, and were as usual busy in destroying the liberties of other nations, their neighbours. Louis the Twelfth (the throat-cutting saint) assisted Julius in reducing Bologna to the most abhorred as the most despicable of all tyrannies—the Papal; and the cup of misery of Italy was as usual full—filled to overflowing by Celtic protection. A Protestant prince of Orange commanded the French troops merely, I suppose, by way of amusement, with a chance of plunder, and the keeping his hand in use at these sort of things; and it might have been through him that the celebrated sketch-book of Da Vinci found its way to Holland, and by William the Third to England. But be this as it may, the siege of Florence was undertaken, and it was on this occasion that Angelo acted as an engineer. His abilities were great, no doubt, but it is lamentable to reflect that a genius of this kind came to be employed in devices so vile as the inventing machinery to batter out the brains of the furious Celtic wild beasts led into Italy for plunder and devastation.

He died on the 17th of Feb., 1563, and was interred at Florence. The vault was opened in 1720, when it was found that the remains had lost but little of their original form.

Before considering the effects which this great man produced on Art during and subsequent to his time, I shall offer a few remarks on his habits and studies. All men of genius are peculiar in their habits; these appear eccentric to the busy practical world who do not understand them nor see their object. They measure genius by a rule wholly inapplicable to it. This is their error. Genius is rare, excessively rare. Select, for example, the form of genius constituting the poet. How many Homers have you? How many lyric poets? Pindar, Sappho, Anacreon, Horace, Burns; that is all, since the beginning of man's career in literature. You must not mingle up with such names your Lake and Cockney poets, nor those of the satanic and demoniac school, of which Byron stands at the head, and old Rantipole Wilson at the other end.

How many genuine critics have you?

Tacitus, Voltaire, Gibbon, Niebuhr? How many dramatists?—one—Shakspeare; unless you add the incomparable Molière, whose path was limited.

Thus you must observe how rare true genius is, and how difficult to measure. The three great men whose lives I now consider, and whose works I now examine, were men of the highest genius. I may venture to draw a comparative parallel of their thoughts and works (as they all laboured on Art) towards the close of this memoir. In the meantime permit me to make a few additional observations respecting Angelo, and proceed next to consider the nature and character of his works.

He died on the 17th of Feb., 1563, at an advanced age. His habits were quite retired—his acquaintances select. He studied with Realdo Colombo, a surgeon, and very good anatomist; and he is thought to have studied anatomy deeply. Vesalius was in the field and had rendered anatomy, no doubt, popular with men of high attainments. Still I do not think that Angelo's anatomical knowledge extended much beyond the surface. He studied, like Da Vinci, every description of science. Dante was with him a favourite author: also the Old and New Testaments.

Of anatomy, Condivi distinctly says that Angelo's knowledge was confined to those structures, which chiefly interest the artist. Like Leonardo he completed few works in sculpture, but this may be ascribed partly to circumstances beyond his controul. He held, with all great artists, that the painting which merely imitates the visible appearances of bodies is not superior to the production of any mechanical trade or employment.

On architecture he held peculiar opinions, borrowed probably from an obscure passage in Vitruvius.

Mr. Duppa remarks in his "Life of Angelo" that at that period the style called picturesque had not been invented. It is much to be regretted that it ever was.

Though a sculptor, the painting of the

ceiling of the "Sistine Chapel" is Michael Angelo's greatest work; and on the character of this and his other works I beg to offer a few concluding remarks.

There is a copy of the "Last Judgment" by Angelo in the School of the Fine Arts in Paris, and of the size of the original fresco painting, which I never saw; but even from this copy it is easy to see that in drawing the human figure, in grandeur of composition, and in the grouping of masses of men, no artist who ever lived could excel Michael Angelo. He regretted, towards the close of life, that he had paid too little attention to grace and beauty in the female form, as developed in the antique statue, thus discovering, though too late, the value of the unapproachable figures of antiquity carved agreeably to the canon of the Greek. Nevertheless, he was at times fully alive to the superiority of the Greek canon compared with the living Italian model before him; and there are some monumental figures, the product of his great hand, in the same School of the Fine Arts to which I have just alluded, an inspection of which will be found fully to bear me out in this opinion.

For many years of his professional career

there cannot be a doubt but his anatomical studies misled him and misdirected his views. We have seen that such studies had no such influence over Leonardo; and Raphael we shall find, knew little or nothing of anatomy. Indeed, in many of his sketches and drawings he seems to have forgotten altogether that Nature had bestowed on man an envelope, which, besides serving many other important purposes, and gifted therefore with many qualities, bestows on man, and especially on woman, that amount of decoration which the human mind looks for and delights in when found.

Thus, whilst second to no man in power, he yet fell behind Da Vinci and Raphael in expressing the graceful and the beautiful, whilst even to manly forms he was but too apt to give an excess of muscular force but seldom displayed in the efforts of living men.

Of the superiority of the antique he must have been himself fully sensible. His attempt to restore the hand of the "Apollo Belvidere" must have taught him this. Misled by the Latin version of the Hebrew Scriptures, he has sculptured Moses with visible horns on his head. It were well to remove them, and to correct, by a new edition, the philological error of the Romish church.

It is recorded of the cartoon drawn by him in the Palazza Vecchio, that it was wonderful, and astonished all who beheld it. The cartoon itself is now lost, but there exists, I believe, an engraving of it.

In loftiness of thought, originality, powers of composition, correctness of outline, no man excelled Angelo after he had arrived at mature years. The patronage extended to him interfered with, rather than assisted, his genius.

In the life and labours of Michael Angelo, we have an instance of the misapplication of Science to Art. He studied anatomy, and for a long time misunderstood its true relation to Art. This grand error he partly corrected towards the close of life, but it is doubtful if he ever wholly overcame it. Nor can I discover that his anatomical studies were ever of that deep, exact, and truthful character which characterized all Leonardo's labours.

The true relation of Descriptive Anatomy to Art was misunderstood, then, by Michael Angelo and his school. In the life of the illustrious and immortal artist which is to follow, and with which I shall conclude this brief work, we shall find nothing in contradiction with the principles here laid down. Raphael's path was his own. He was the painter of men as they show themselves under the influence of mind. His own mind was the emblem of truth, and of the sublimest generalizations.

SECTION III.

Raphael.

THIS prodigy, for such he assuredly was, was born at Urbino, in the Papal states, on the 28th of March, 1483. No country but Italy, it would seem, can produce such men. At the time of his birth Angelo was nine years of age, and Leonardo forty, or thirty-one, as the case may be. When Raphael was twenty years of age Leonardo was sixty, or at least fifty-one, and Angelo twenty-nine. When Raphael was twenty, these great masters stood thus :—

Leonardo . .	60 or 51
Angelo . . .	29
Raphael . .	20

His original name was De Santi, which was changed in time to Sanzio. His father was an artist of some local reputation. He placed his son, the immortal Raphael, with Perugino. Such was the educability of his mind, and the pliability and facility of hand, that already, when a mere lad, his works, and those of his master, seemed identical. This reacted even on his master, who seemed to improve after Raphael became his student.

It seemed to have been Raphael's destiny to influence the world of Art more than his illustrious cotemporaries, or perhaps any that ever lived. His knowledge of form, of proportions, and his perception of truth were absolutely perfect. He had seen the antique marbles, and likewise the works of Leonardo and Angelo. He encountered no difficulties in life, and thus every advantage being heaped upon him, he rose at once to the summit of reputation.

I have already mentioned the disinterment of the antique statues, and the vast influence this discovery exercised over Art. It enabled the illustrious trio to reach at once the highest point of perfection. Italy possessed no such living forms. This Raphael knew well. But he saw that if he merely copied the antique, as so many have done, from Carlo Dolci downwards, he could lay no claim to originality. His genius led him thus to look into Nature once more, and to select that form, and that beauty which the antique disregarded—the forms of ordinary life, and the beauty of expression. To this he added the highest powers of composition.

This then was the plan of Raphael's studies. The Greek he knew and deeply observed; the living forms before him he drew from Nature; his drawings and his portraits speak a language not to be mistaken. Never did artist succeed more perfectly in animating the canvas. Of anatomy he knew nothing, and must have been quite sensible of the misdirection of Angelo's studies. Whilst with other artists a life-like appearance is held to be of importance, with Raphael it was everything. At seventeen he made his first great public effort, his master having been called away for a time from Florence. To the manner of his master he added grace, and a certain amount of beauty of form. He next repairs to Florence, and in 1506 to Rome.

As the Cartoon of Pisa, drawn by Angelo and Leonardo, emancipated at once the artistic mind from all its fetters, so those of Raphael preserve, and will preserve, Art for ever from retrograding greatly. Raphael never copied Angelo; he never copied any one, but he saw at a glance the grandeur of Angelo's frescos, and he aimed at equalling or surpassing that grandeur. That he equalled it there cannot be a doubt. At twenty-four he was in the middle of his career, and desirous of measuring his strength with Angelo. He obtained, through a distant relative of his, an engagement from the Pope to repaint the Vatican. Here he painted his first grand works. It excites our wonder to find that in "his School of Athens," he created, by the force of genius, classic heads and costumes which have subsequently stood the test of archaiological research and discoveries.*

Raphael is said to have incessantly studied the antique. De Quincy conjectures that from this he was led to represent men in thought, whilst Angelo's anatomical studies led him to represent men in action. His highest composition is usually considered to be "The Miracle of Bolsena;" "The Heliodorus" is thought to have no equal.

He remained twelve years at Rome, when,

* That genius can do much, I admit; but this anecdote I doubt; and feel disposed to ascribe Raphael's knowledge of the costume of Ancient Greece to access to works which even then must have abounded in the valuable libraries of Italy.

by force of circumstances, he became an architect; built the Loggia of the Vatican, and revived the arabesque, or grotesque—so called from the ancient remains of this style having been found in grottoes. He owed his ideas on this point to the paintings he saw in the Baths of Titus, then in full perfection. He was assisted in these drawings by Giovanni de Urbino. They are beyond what we term Nature; although it may be admitted that, as respects the forms of animal life, recent geological discoveries have made it difficult to say what is, and what is not, beyond Nature. In such hands as Raphael's and Rubens', the arabesque and the allegory succeed; in those of Fuseli, Martin, and others, they become ludicrous. In the works of great minds, all anachronisms and incongruities are overlooked. They can handle the gigantesque, the arabesque, the outrageous, without offending. Rubens' Venuses and Junos are very odd-looking persons; they resemble what, no doubt they were, portraits of the common fish-women of Rotterdam and Amsterdam. His Sabine women are unmistakably the large Dutch vrows and their daughters, keepers of pot-houses at the Hagen; but in looking at the

"Rape of the Sabine Women," the observer cares nothing for these anachronisms in time and place. But let a man of mediocrity try all this, and see what a torrent of well-deserved criticism would be poured out on him. "He has violated established conventionalities," says one; "his anachronisms," says another, "betray astounding ignorance." But before the pencil of the great master, the world bows down, and simply admires. Men feel themselves in the presence of a work which they are couscious no man now can equal, or even approach. This is the whole secret. In drawing man you must not forget his *intellectual* mind, although it is his instinctive mind which no doubt it behoves the artist most to study. What constitutes the real difference in the merits of portraits? It is this: one artist seizes on the expression which has been bestowed on the features by the instinctive mind of the individual; another, not understanding this, draws the man with features arranged by reflection. Now men of deep reflection are rare, and, therefore, this process of thought seldom leaves unalterable traces on the face. The passions also tell strongly on the features, and these are beyond controul. The great artist looks through the material mask and reads the nature, that is, the truth, which lies beyond it. His portraits, therefore, have been true to Nature, are likenesses of the man and of his mind, whatever that may be.

Raphael's incessant study of the antique enabled him unquestionably to have a perfect *form* in his mind by which he measured all other forms. In the antique sculptures of Greece, he with his great rivals, saw the only *real;* and the Madonnas of Guido are copied from it. When De Quincy says that Guido's Madonnas are Pagan, he does so, I presume, from the idea that they are copies of the Greek. But the remark, though coming from a most ingenious man, is not at all applicable, whichever of the theories you adopt. The modern Italian woman, with her heavy elongated features, dull eyes, flat head, and sombre look, has no more of a Christian look than the glorious busts of antiquity. Nor am I aware that Christianity affects or alters the human countenance in the slightest degree. Modern Christians are usually clothed, but ancient Pagans were perfectly well off in this respect.

As to Raphael inventing an ideal head for

his Madonnas, I need not say one word. Raphael knew perfectly that men cannot *imagine fine forms*; they must be seen in Nature before they appear in Art. From the commencement of his career, as it seems to me, Raphael endeavoured constantly to represent the minds and passions of men and women; to paint them not only alive, but as influenced by the nature of their minds; by their present thoughts and prevailing emotions. This was his scheme—this his secret. He selected the best formed of his countrywomen he could find, and by exaggerating some points, and diminishing others, he caused the face somewhat to approach the perfect, or antique, though with features differently grouped and proportioned. He was the painter of expression, that is, of the mind and its emotions. His own mind was delicate and tender. He seems to have been the great improver, if not the inventor, of the art of engraving on copper. He drew expressly for the engraver, and it is even thought that he sketched the outline of the "Massacre of the Innocents" on copper. At all events he seems to have reduced to a science, or principle, that art which represents the harmony of colour in pictures, by a harmony of lines. Had he lived longer he would also have become a sculptor. As a portrait painter he stands unrivalled. He painted in his portraits the inward man.

So great and so able an artist naturally drew around him a school of great artists; artists who caught up his spirit and style. Andrea del Sarto drew portraits which were mistaken for those of his illustrious master. After three hundred years, Raphael's portraits seem still alive.

Every particular in the life of this great man, serving to explain his method and style, must be interesting. It seems that for some time, at least, before his death, it was his habit to compose and design or sketch some great work. Julio Romano took up the work of the picture, that is, painted it, and Raphael retouched and finished copies made by his pupils. This explains, to a certain extent, the number of works which pass under his name. The method is not applicable to science.

His knowledge of the nude figure was not derived from anatomy, but from the study of living forms.

In 1512 he seems to have superseded

Angelo who was sent to the quarries. In 1515, at the death of Bramanti, he was appointed architect of St. Peter's. Here a new difficulty occurred. His predecessor, Bramanti, incredible as it may appear, had been working without a plan. This extraordinary deficiency Raphael also overcame by the force of genius. He found time to decorate the house of his friend Chigi with paintings from the romance of Apuleius, the earliest of all romances. The Cartoons of Hampton Court were drawn by his master-hand. They are beyond all praise.

The "Battle of Constantine" was also his, and many other immortal works. The "Transfiguration," was his last great work.

On the 7th of April, 1520, at the early age of thirty-seven, the career of this unsurpassed genius was suddenly cut short by death, leaving, however, a reputation which must endure for ever

Adequately to represent in poetry or on canvas the deeds of great men requires a corresponding genius. Achilles owed much to Homer. This great truth flashed strongly on my mind whilst looking at the modern paintings in Versailles, commemorative of the mighty deeds of Napoleon. The paintings are good enough, but when weighed in the scale against his great name, are obviously deficient. In like manner some of the greatest naval deeds of British men have been rendered ludicrous and vulgar by the vile and truly abominable trash of Dibdin's songs. If you compare Campbell's beautiful Odes with the vulgar productions of the Saxon Boor you will see immediately what I mean.

Literature and art must thrive or sink together. A Satanic school of poetry must give rise to a Satanic school of art. Under a somewhat modified form, the Dibdin school is still in full force in England; and the punsters and the comic men, the roaring boisterous crew, once more give their concerts with success; "Poll of Wapping," and "Sall of Dover," and the "Old Arethusa," threaten again to drive good taste and high feeling from our theatres. If poetry, lyric poetry, and music be fine arts, they wear at this moment in England singularly vulgar habiliments.

It was in the library at Sienna where the great genius of Raphael first showed itself. In 1503 he first reached Florence, and from this time, as is said, his style altered at once, became bolder, more enobled and perfect; the result, no doubt, of an inspection of Leonardo's and Angelo's works. In facility and rapidity of execution he probably exceeded all men.

In 1504 he first saw the remains of the Antique, in the Medici Palace; and from that time continued more and more to add beauty of form to grace and expression. The Cartoons now in Hampton Court, as specimens of drawing, leave nothing to be desired.

In this way he contrived to give grace and even beauty to the ordinary Italian woman, who generally has neither. His Madonnas are Italian women modified by his knowledge of the Antique. He did not invent the Arabesque or grotesque, but copied it from the drawings in the Baths of Titus, which he is said to have destroyed after copying them. But this insinuation against his character has been refuted. His mind was noble and simple, and free from guile. He employed the Arabesque style merely as allegorical.

By many he is thought to be the very first of all painters, in respect of composition; but I do not think him superior in this respect to Leonardo and Angelo.

Guido Rheni is supposed to excel Raphael and all others in painting the Madonna; and unquestionably no more exquisite and beautiful face was ever placed on canvas, than the Madonnas of Guido. Weighing attentively the subject I ventured to come to this conclusion, namely, that Guido Rheni having seen the Antique Marbles, perceived in them the perfection of the human form; beyond which no man could proceed. Acting on this idea he painted those divine faces, with glorious flesh tints, blue eyes, and flowing flaxen hair. But as there really are no such persons in Italy it was natural for De Quincy and others to imagine that Guido Rheni had invented them, and that, like the Antique, as they supposed, his Madonnas merely represent the ideal. But I have explained elsewhere that there exists no such thing as *ideal* in any sense of the *term.**

I find, on examining the lives of other celebrated artists, that some studied the superficial anatomy of man, whilst others, equally cele-

* My esteemed friend Dr. A. Sutherland informs me, that in a valley in the Tyrol, he met with a number of women, who, in complexion and form of features, bore a striking resemblance to the Madonnas of Guido.

brated, neglected this study altogether, or were content with the knowledge they obtained from engravings, drawings, or from the *écorchée* figure. Poussin commenced his anatomical studies in Paris, and continued them under a distinguished surgeon, Nicholas Lascke. At Rome he began a new course of practical anatomy and read Vesalius, making extracts as he read. After this he took to the living models. Like many men of genius, Poussin suffered from the neglect of the rich and powerful, all such having abandoned him for some Court parasite and favourite, now forgotten. A countryman of his, one Jean Daghet, a cook, cherished and supported him. Daghet's son, Gaspar, took Poussin's name, and became a good artist. Patronage seems to me fatal to genius; it is almost always exercised against it. When I look at the engravings of some Court painters of the present day, and compare them with the remains of the great masters, the mind is filled with astonishment. It is this which has induced so many, as well as myself, to endeavour to discover a cause and to assign a reason for a fact which cannot be denied, and which cannot be concealed; namely, the vast superiority of the ancient masters over modern artists. We have seen that the theory which ascribes to a deep knowledge of anatomy the superiority of the great Italian masters is false, in fact, even admitting the utility of a study well calculated, no doubt, to correct the eye of the artist when defective in observing powers. In all such cases a knowledge of anatomy must be useful.

CONCLUSION.

The Fine Arts, and more especially the divine compositions of the sculptor and the painter, do not in any way contribute to the political power or wealth of a country; but united with literature and science, as they generally are, they mark the position a nation and a race are to hold on the page of history. Driven from all universities and schools of high education; unheeded by the Church; its professors held of low repute by the aristocratic and commercial wealthy, I do not see how the Fine Arts can thrive in Britain.

Writing respecting the condition of the arts in the time of Diocletian, Gibbon remarks:—"If such was, indeed, the state of architecture, we must naturally believe that painting and sculpture had experienced a still more sensible decay. The practice of architecture is directed by a few general and even mechanical rules. But sculpture, and, above all, painting, propose to themselves the imitation not only of the forms of Nature, but of the characters and passions of the human soul. In those sublime arts the dexterity of the hand is of little avail, unless it is animated by fancy, and guided by the most correct taste and observation." If it be thus that nations are weighed in the balance by the philosophic historian, how will Britain fare?

There is but one school of art—Nature. But, to read her volume profitably, artists must study profoundly the antique Greek, and ancient Italian school, formed by the era of Leonardo, Angelo, and Raphael.

It may precede or follow or coincide with the study of the living figure; still these immortal works must be your guide. For whether it be composition, or colouring, or design, you are likely to find that these masters read Nature more clearly than you ever can. But do not copy nor imitate them further than as objects of study.

Learn anatomy by all means, but do not

forget its object. When you draw a dissected limb be sure to sketch the living one beside it, that you may at once contrast them and note the differences. In drawing from the nude figure, contrast your sketch with the antique; you will find in it many defects. Never forget that perfection, the result of a high specialization of Nature's law of individuality, is rare; the opposite, that is, imperfection, the result of a tendency to unity of organization, is by far the more common. You will be chiefly called on to draw the draped figure: see that you place your drapery not on a machine but on a person of *fine feeling*. Fashion in dress is the trick of society, to substitute a conventionalism for beauty and fine forms; never sacrifice art at its shrine, but paint the person in what *becomes* him or her, regardless of the existing mode.

The relation Anatomy holds to Art is to explain—first, how far the shapes and figures of the inward structures modify the external forms of man and woman;—second, it informs the artist of the meaning of such forms;—third, it explains to him the laws of deformation; that is, of variety in external forms; the causes of these varieties, and the tendency to which they lead. As an artist he must represent them, no doubt; but in doing so let him wisely follow Nature rather in her intentions than her forthcomings, and return to the perfect or to its approximation, whenever time and circumstances permit him to do so.

CUVIER.

NOTE I., p. 17.

IN 1815, the exact nature and value of true descriptive anatomy were unknown in England; the great text-book of the English school was "The London Dissector." What Cuvier called comparative anatomy, but which I have shown to be merely the exact descriptive anatomy of species, and through them of genera, continued to be unknown in England so late as 1825. Its very object was wholly misapprehended. Hunter understood it perfectly, but in his vast researches he had other objects in view. These were chiefly physiological or the laws of life. Still his researches proved the non-essentiality of forms in the grand scheme of Nature. The museums of that period (excepting his own) were mere collections; the British Museum, as it was called, was a mere chaos. In 1822 there was not a skeleton of a fish, or a perfect one of any mammal in the museum of the College of Surgeons of London; a few fossil bones had been collected by accident. It were well for the members of the Corporation that no such collection had ever existed, and thus two hundred thousand pounds, expended on what must ever appear foreign to the medical and surgical profession, might have been turned to a better account than in vain endeavours to acquire for a corporate body of surgeons a scientific character which it can never attain. I have sometimes thought it doubtful if the true character of descriptive anatomy be yet understood in Britain. It takes a century, or more, before the great original ideas of a man like Bichat, a scientific thought of vast import, can make its way amongst a race, antagonistic of new thoughts and of all science which leads not directly and immediately to a profitable end. To induce the student to remember any anatomical fact, the teacher, I remember, was obliged to couple it with another, to show the profit likely to accrue to him by remembering it.

Note II., p. 22.

When Cuvier visited commercial England, he discovered the only specimen of the cranium of the Balæna Mysticetus, the whale of commerce, at that time in Europe, in a dark vaulted cellar below the British Museum; unheeded, unknown, and covered with soot and dust. Beside it lay the cranium of the South-sea whale. I found them many years afterwards in the same place. Zoological science could descend no lower. Matters are not improved.

NOTE III., p. 27.

The pseudo-scientific cliques of Britain made, at first, a determined stand against transcendentalism in anatomy and the doctrines of unity of the organization. Their position as educational employées and officials necessitated this. This resistance to science, however, could not continue, and some of the party gradually slid into an unobtrusive low-transcendentalism, in hopes of deprecating the scrutiny of "the powers that be," yet claiming for themselves a wish rather expressed than understood, not to be thought some hundred years behind continental science. This concession, however, has not been fully granted, even yet, and the reins are held by a tight hand; but no doubt hopes are entertained that it will be conceded, seeing the stout fight they now maintain with the transitionalists,—meaning the school of Buffon and Geoffroy. As usual, they commence by misrepresenting the doctrine. Many animals, say they, the pterodactile, for example, stand alone and apart from all others, and from the epoch to which they belong. Now this, in the first place, I must deny. But supposing it to be the case, what has this to do with the appearance of forms having but slender relations seemingly to the Fauna of the Epoch! We know not why the series or chain is interrupted; we know not why it is resumed, simply because the secondary laws effecting these great revolutions in the form of life, have never yet been properly studied. The Ornithorynchus and the Pterodactyle, Chirotherium and the Dinorbis, the fossil of the Cape, will find their true place in the scheme of living forms by-and-by. Of the past you but know a mere fragment; the fragment discovered by a man whom I had the pleasure to count amongst my friends. The future is yet to come; its organic forms you can no more conjecture than you could the past, until Cuvier disinterred them from the quarries of Montmartre. Do not mistake your position, and the resistance you have offered, reluctantly, I admit, to the progress of organic science. The mingling of your names with theirs will not succeed in deceiving posterity as to your real views. You resisted the Theory of the Unity of Organization, until you became ashamed of longer opposing the Doctrine.

NOTE IV., p. 37.

With M. Flourens, "Vestiges of Embryonic Structure" have no meaning,—"gradual and successive formations of higher orders of life" point to nothing,—and he scrupulously avoids noticing the late researches and discoveries of De Blainville, tending to destroy all idea of species and genera, and to show that species mark an epoch in time, and not a distinct animal. To this conclusion, Geoffroy and Humboldt had long before arrived.

NOTE V., p. 40.

The errors of naturalists in respect of fossil and other bones, found in strata, whether alluvial, diluvial, or of more ancient date, continued long after the publication of the "Ossemens Fossiles." I was a member of a society whose object was chiefly geological and mineralogical. Attending one day in my place as a member, I listened to the reading of a paper on certain bones which had been found at a very considerable depth below the surface, between Loch Lomond and the River Clyde. The bones were described in the Memoir, as resembling those of a fox. Now on the table before me there lay some of the bones of *a seal*. When the reader had concluded his paper, the chairman suggested that my opinion should be taken in respect of the bones in discussion; they had never thought of this before. The remarks I made to the society were very brief. I said "that surely the paper just read could not have reference to the bones on the table before me, inasmuch as these bones in no way resembled the bones of foxes, but had belonged to a *seal*." The society, taken by surprise, took measures to erase the awkward occurrence from their minute-book, and to read the Memoir so corrected, subsequently.

The principle discovered by Bichat and Cuvier, the new element of science applied by them to the animal world, was at that time, and I feel disposed to think, even now, all but unknown to British Naturalists. Yet there can be nothing more simple; it is this, perhaps, which renders its comprehension difficult.

GEOFFROY.

NOTE I., p. 55.

It was excellently observed by the talented MacCulloch (author of the "Remarks on the Islands and Highlands of Scotland"), that the uneducated solve every difficult moral, physical, or metaphysical problem, by an appeal at once to a First Cause. The practice prevails also with the very learned—in words—as at Oxford and Cambridge, and thus extremes meet.

NOTE II., p. 95.

I met Le Vaillant in Paris in 1821-22; he is since dead. He was not a scientific naturalist, for there were none in his days, but a man of great powers of observation; a lover of nature; shrewd enough, but it seemed to me without education. Yet he must have written the works which go by his name. He was vain, and quite a Frenchman. There was no foundation for what was said of him by Barrow. He crossed the Orange River, and was encamped for a long time on the banks of the Little and the Great Fish River, close to the spot where I lived for some years. His works may be entirely depended on.

Le Vaillant's great object on his return to France was to dispose of his collection, in which he succeeded. I never heard his name mentioned by any one at the Garden of Plants, excepting by M. Roger, who introduced me to him. He seemed a man of a by-gone age.

NOTE III., p. 95.

If the definition usually given to species be false, the whole scaffolding of natural history, as it now exists, in genera, natural families, groups, must also be false. They fall together; nevertheless, so long as the causes of the multitudinous forms which people the earth, the air, and the waters, be unknown, and species retain a permanency of form during historic periods of long duration, of ages, in fact,—so long will the permanency of species remain a fixed idea with men, though philosophically proved to be incorrect. The earth and planets move round the sun, but the sun's course through the heavens is still the ordinary language of men. With men the sun rises, and the sun sets, notwithstanding the demonstrations of the divine Newton proving these notions to be false, and the language incorrect. The influence, then, of the great truths of philosophy is, to a certain extent, limited by the circumscribed nature of Human Thought.

NOTE IV., p. 96.

In the "Life of Etienne Geoffroy," by his son Isidore, I find by far the best account of Cuvier. It was the discovery of the *principle* which immortalized Cuvier. The repetition of descriptions of new fossil species presenting some few peculiarities, is simply nauseating, and below notice. It is with this as it was with the so-called comparative anatomy. The principle being discovered by Cuvier, it was in vain that others endeavoured to keep up men's interest in the subject. Meckel's great work excited not the smallest attention. Men had lost all interest in the anatomy of dogs and cats, spiders and shell-fish. To talk of the fossil zoology of Britain is pure nonsense; for, in the first place, Britain, in the times alluded to, bore no resemblance to what it is now. We do not even know that it was an island; and, secondly, there can be no true zoology in the absence of all, or nearly all, the external characters.

NOTE V., p. 115.

One Bernard Pallissey, a potter mentioned by Fontenelle, was the first to call the *Lusus Naturæ* theory into question. In 1762, Daubenton first proposed the question of fossil remains, in a scientific manner; he was followed by Pallas in 1792. As early as 1670, however, Augustin Scilla restated forcibly the observations of Pallissey: Leibnitz in 1603; Buffon followed. But he confined himself to the idea that a single species of large quadrupeds had been lost; Cuvier appeared, and boldly stated, that all the then-existing species of animals had been destroyed; and although we know now, that a statement of this kind is far from exact, the proposition remains a colossus in dignity and strength.

On the 1st Pluviose, an. IV., a mode of dating I am thankful not to understand, Cuvier read his first great memoir to the Institute. It then met for the first time after its republican organization.

NOTE VI., p. 132.

To Cuvier's theory of the "Fixity of Species," as demonstrated by the drawings on the Egyptian Tombs, Geoffroy objected, that "as the surrounding circumstances had not changed, there existed no reason for any change in the Fauna." He might have added, that the period referred to by Cuvier, in proof of his views, was but an instant in the duration of the globe.

Convinced of the soundness of the basis on which Autenrieth, Goethe, and Geoffroy had constructed the great theories of Transcendental Anatomy, I hesitated not applying them constantly in all my researches in zoology, from

1820 inclusive: these principles were fully explained by me in three courses of lectures on Comparative Anatomy delivered to distinguished classes in 1825–26–27.

At first the doctrines were held by most persons to be entirely theoretical, fanciful, and of no value practically. I have lived to see such ideas abandoned by those who maintained them. I have myself ever found the transcendental doctrines eminently practical. They solve, as high generalizations must ever do, many difficult and puzzling questions in Human and Comparative Zoology. Trusting to it as to a sure guide, I was led to observe the presence of rudimentary teeth imbedded in the upper jaw of the fœtus of the Greenland whale, placed above the whalebone. It alone explains the true philosophy of the teeth, and in it will be found the explanation of some of those difficult questions which have been raised in respect of the natural history of the salmon, and especially of one, namely, the identity of the samlet, or parr, with the smolt, or young, salmon. By the law of unity of the organization, the salmon-fry, whilst confined within its egg-coverings, and for some time after its escape into the waters, exhibits certain characters of form and colouring which do not belong to it especially *as a salmon*, but which it has in common, 1st, more especially with animals of its own kind—the genus salmo. — 2nd, peculiarities of form which it has, as an embryo, in common with the now-existing race of fishes, and with all that has lived. These embryonic forms and colouring the young salmon throws off in time, as it becomes developed, but not all at once. Thus the elongated dorsal fin which the embryo salmon has, in common with the fishes of primitive times, disappears early; but the dark markings on the sides—generic characters which it has in common with many, if not all the salmo-tribe—it retains for some time after its escape from the egg. In course of time, as the smolt grows, it lays aside all, or nearly all (for a few dark spots are still found, even in the adult salmon, on the sides, *above* the lateral line) its embryonic forms and colouring, assuming its specific characters.

This is the history of all that lives; and it forms a chapter in the history of the error of those who mistake a brandling parr, or samlet, for a young salmon; who, ignorant of science, mistake a transitional, and generic, and embryonic colouring for a specific character of the animal. The parr markings on the sides of the salmon fry no more prove the identity of the salmon and parr, than the presence of the elongated dorsal fin its identity with the fossil fishes of unknown antiquity; or the webbed fingers of embryo man the identity of man with the seal. They represent simply the generic, not the specific, characters. The parr is not a young salmon. This view I maintained thirty years ago, and have not yet seen any reason for departing from it. Seemingly, without being aware of it, Cuvier constantly employed the transcendental principles in the last edition of his great work on Fossil Remains. His elaborated inquiry into the composition of the osseous head of the crocodile, is simply an effort to determine the homologous bones of reptiles and mammals.

The application of the transcendental in anatomy to art, is quite as practical as to zoological and geological science; it alone explains the variety in normal forms in harmony with Nature; and of anormal forms antagonistic of the existing order of things.

THE END.

LONDON :
Printed by SAMUEL BENTLEY & Co.,
Bangor House, Shoe Lane.

Printed in the United States
By Bookmasters